U.S. Fire *Administration*/*National Fire Academy*

Field Operations Guide

ICS 420-1

July 2010

Ten Standard Fire Orders

FIRE BEHAVIOR

a. Keep informed on fire weather conditions and forecasts.
b. Know what your fire is doing at all times.
c. Base all actions on current and expected behavior of the fire.

FIRELINE SAFETY

d. Identify escape routes and safety zones and make them known.
e. Post lookouts when there is possible danger.
f. Be alert. Keep calm. Think clearly. Act decisively.

ORGANIZATIONAL CONTROL

g. Maintain prompt communication with your forces, your supervisor, and adjoining forces.
h. Give clear instructions and ensure they are understood.
i. Maintain control of your forces at all times.

IF YOU CONSIDERED 1 THROUGH 9, THEN

j. Fight fire aggressively, having provided for safety first.

Common Denominators of Fire Behavior on Tragedy Fires

- Most incidents happen on the smaller fires or on isolated portions of larger fires.
- Most fires are innocent in appearance before the "flare-ups" or "blow-ups." In some cases, tragedies occur in the mop-up stage.
- Flare-ups generally occur in deceptively light fuels.
- Fires run uphill surprisingly fast in chimneys, gullies, and on steep slopes.
- Some suppression tools, such as helicopters or air tankers, can adversely affect fire behavior. The blasts of air from low flying helicopters and air tankers have been known to cause flare-ups.

U.S. Fire Administration

Mission Statement

We provide National leadership to foster a solid foundation for our fire and emergency services stakeholders in prevention, preparedness, and response.

STATEMENT OF INTENT

The content of the Field Operations Guide (FOG) is intended to provide guidance for the application of the Incident Command System (ICS) to any planned or unplanned event. Position descriptions, checklists, and diagrams are provided to facilitate that guidance. The information contained in this document is intended to enhance the user's experience, training, and knowledge in the application of the Incident Command System. All users must obtain proper ICS training at the level necessary to effectively utilize the system.

TABLE OF CONTENTS

i

CHAPTER 1

COMMON RESPONSIBILITIES

COMMON RESPONSIBILITIES

The following is a checklist applicable to all ICS personnel:

a. Receive assignment from your agency, including:
    1. Job assignment, e.g., Strike Team designation, overhead position, etc.
    2. Resource order number and request number
    3. Reporting location
    4. Reporting time
    5. Travel instructions
    6. Any special communications instructions, e.g., travel frequency
b. Upon arrival at the incident, check in at designated Check-in location. Check-in may be found at:
    1. Incident Command Post
    2. Base or Camps
    3. Staging Areas
    4. Helibases
    5. If you are instructed to report directly to a line assignment, check in with the Division/Group Supervisor.
c. Receive briefing from immediate supervisor.
d. Acquire work materials.
e. Conduct all tasks in a manner that ensures safety and welfare of you and your co-workers utilizing accepted risk analysis methods.
f. Organize and brief subordinates.
g. Know the assigned frequency (ies) for your area of responsibility and ensure that communication equipment is operating properly.
h. Use clear text and ICS terminology (no codes) in all radio communications. All radio communications to the Incident Communications Center will be addressed: "(Incident Name) Communications," e.g., "Webb Communications".

i. Complete forms and reports required of the assigned position and send through supervisor to Documentation Unit.
j. Respond to demobilization orders and brief subordinates regarding demobilization.

UNIT LEADER RESPONSIBILITIES

A number of the Unit Leader responsibilities are common to all units in all parts of the organization. Common responsibilities of Unit Leaders are listed below. These will not be repeated in Unit Leader Position Checklists in subsequent chapters:

a. Participate in incident planning meetings as required.
b. Determine current status of unit activities.
c. Confirm dispatch and estimated time of arrival of staff and supplies.
d. Assign specific duties to staff and supervise staff.
e. Develop and implement accountability, safety, security, and risk management measures for personnel and resources.
f. Supervise demobilization of unit, including storage of supplies.
g. Provide Supply Unit Leader with a list of supplies to be replenished.
h. Maintain unit records, including Unit/Activity Log (ICS Form 214).

Notes

CHAPTER 2

MULTI-AGENCY COORDINATION SYSTEM (MACS)

Contents

MULTI-AGENCY COORDINATION SYSTEM (MACS)

A Multi-Agency Coordination System (MACS) is a combination of facilities, equipment, personnel, procedures, and communications integrated into a common system with responsibility for coordination of assisting agency resources and support to agency emergency operations.

MACS FUNCTIONS

a. Evaluate new incidents.
b. Prioritize incidents:
    Life threatening situation
    Real property threatened
    High damage potential
    Incident complexity
c. Ensure agency resource situation status is current.
d. Determine specific incident and agency resource requirements.
e. Determine agency resource availability for out-of-jurisdiction assignment at this time.
f. Determine need and designate regional mobilization centers.
g. Allocate resources to incidents based on priorities.
h. Anticipate future agency/regional resource needs.
i. Communicate MACS "decisions" back to agencies/incidents.
j. Review policies/agreements for regional resource allocations.
k. Review need for other agencies involvement in MACS.
l. Provide necessary liaison with other coordinating facilities and agencies as appropriate.

POSITION CHECKLISTS

MAC GROUP COORDINATOR - The MCCO serves as a facilitator in organizing and accomplishing the mission, goals and direction of the MAC Group. The Coordinator will:

a. Facilitate the MAC Group decision process by obtaining, developing and displaying situation information.
b. Activate and supervise necessary unit and support positions within the MAC Group.
c. Acquire and manage facilities and equipment necessary to carry out the MAC Group functions.
d. Implement the decisions made by the MAC Group.

MAC GROUP AGENCY REPRESENTATIVES - The MAC Group is made up of top management personnel from responsible agencies/jurisdictions, those organizations heavily supporting the effort or those that are significantly impacted by use of local resources. MACS Agency Representatives involved in a MAC Group must be fully authorized to represent their agency. Their functions can include the following:

a. Ensure that current situation and resource status is provided by their agency.
b. Prioritize incidents by an agreed upon set of criteria.
c. Determine specific resource requirements by agency.
d. Determine resource availability for out-of-jurisdiction assignments and the need to provide resources in Mobilization Centers.
e. As needed, designate area or regional mobilization and demobilization centers within their jurisdictions.
f. Collectively allocate scarce, limited resources to incidents based on priorities.
g. Anticipate and identify future resource needs.
h. Review and coordinate policies, procedures and agreements as necessary.
i. Consider legal/fiscal implications.

j.   Review need for participation by other agencies.

k.   Provide liaison with other coordinating facilities and agencies as appropriate.

l.   Critique and recommend improvements to MACS and MAC Group operations.

m. Provide personnel cadre and transition to emergency or disaster recovery as necessary.

SITUATION ASSESSMENT UNIT - The Situation Assessment Unit (this is also referred to in some agencies and EOC's as the Intelligence Unit) in a Multi-Agency Coordination Center is responsible for the collection and organization of incident status and situation information. They evaluate, analyze and display information for use by the MAC Group. Functions include the following:

a.   Maintain incident situation status including locations, kinds and sizes of incidents, potential for damage, control problems, and any other significant information regarding each incident.

b.   Maintain information on environmental issues, status of cultural and historic resources, and condition of sensitive populations and areas.

c.   Maintain information on meteorological conditions and forecast conditions that may have an effect on incident operations.

d.   Request/obtain resource status information from the Resources Unit or agency dispatch sources.

e.   Combine, summarize and display data for all incidents according to established criteria.

f.   Collect information on accidents, injuries, deaths and any other significant occurrences.

g.   Develop projections of future incident activity.

RESOURCES UNIT - The Resources Unit, if activated in a Multi-Agency Coordination Center, maintains summary information by agency on critical equipment and personnel committed and available within the MACS area of responsibility. Status is kept on the overall numbers of critical resources rather than on individual units:

a. Maintain current information on the numbers of personnel and major items of equipment committed and/or available for assignment.
b. Identify both essential and excess resources.
c. Provide resource summary information to the Situation Assessment Unit as requested.

INFORMATION UNIT - The Information Unit is designed to provide information regarding the MACS function. The unit will operate an information center to serve the print and broadcast media and other governmental agencies. It may provide summary information from agency/incident information officers and identify local agency sources for additional information to the media and other government agencies. Functions include:

a. Prepare and release summary information to the news media and participating agencies.
b. Assist news media visiting the MACS facility and provide information on its function. Promote inter-agency involvement.
c. Assist in scheduling press conferences and media briefings.
d. Assist in preparing information, materials, etc., when requested by the MAC Group Coordinator.
e. Coordinate with Joint Information Center (JIC) if established.
f. Coordinate all matters related to public affairs (VIP tours, etc.).
g. Act as escort for facilitated agency tours of incident areas, as appropriate.

Notes

CHAPTER 3

AREA COMMAND

Contents 3-1

AREA COMMAND

Area Command is an expansion of the incident command function primarily designed to manage a complex or large incident/event or an area that has multiple incident management organizations assigned. An Area Command may be established at any time that incidents are close enough that oversight is required among incident management organizations to ensure conflicts do not arise.

The function of Area Command is to develop broad objectives for the impacted area and coordinate the development of individual incident objectives and strategies. Additionally, the Area Command will set priorities for the use of critical resources allocated to the incidents assigned.

The organization is normally small with personnel assigned to Command, Planning and Logistics functions. Depending on the complexity of the interface between the incidents, specialists in other areas such as aviation, hazardous materials, environment, and finance may also be assigned to Area Command.

AREA COMMAND ORGANIZATION FOR THREE INCIDENT MANAGEMENT TEAMS

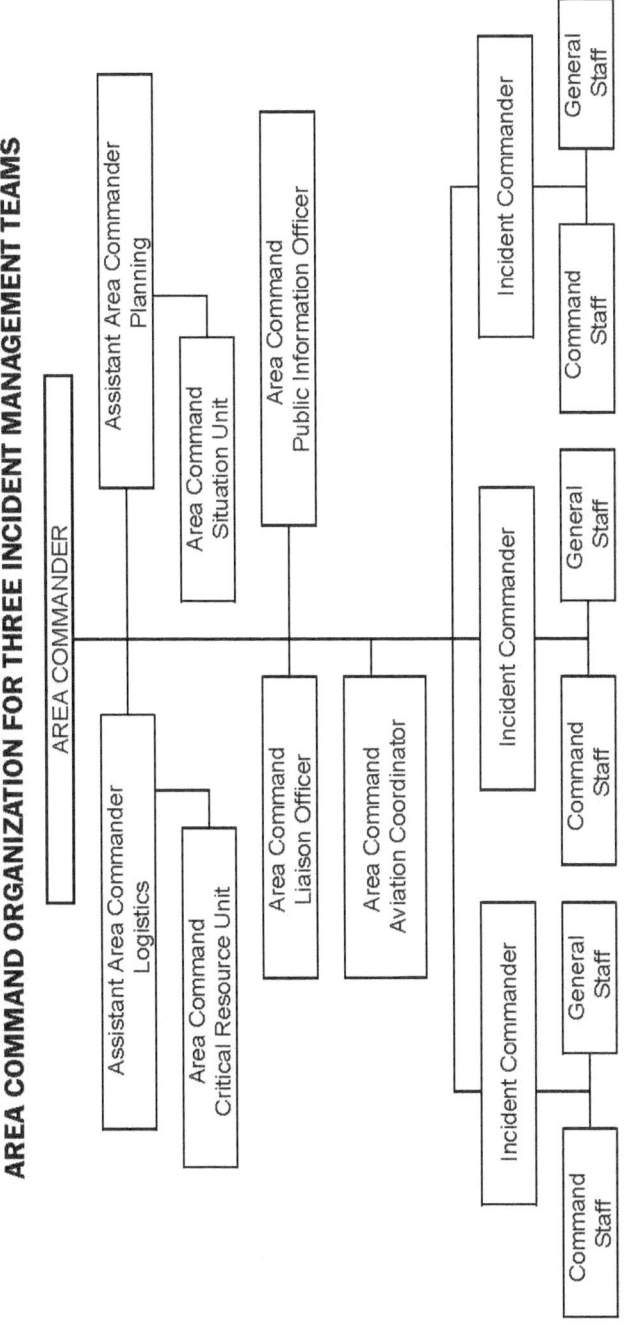

AREA COMMAND

AREA COMMAND

POSITION CHECKLISTS

AREA COMMANDER (Single or Unified Area Command) -
The ACDR is responsible for the overall direction of incident management teams assigned to the same incident or to incidents in close proximity. This responsibility includes ensuring that conflicts are resolved, compatible incident objectives are established and strategies are selected for the use of critical resources.

Area Command also has the responsibility to coordinate with local, state, federal and volunteer organizations and agencies that are operating within the Area:

a. Obtain briefing from the agency executive(s) on agency expectations, concerns and constraints.
b. Obtain and carry out delegation of authority from the agency executive for overall management and direction of the incidents within the designated Area Command.
c. If operating as a Unified Area Command, develop working agreement for how Area Commanders will function together.
d. Delegate authority to Incident Commanders based on agency expectations, concerns and constraints.
e. Establish an Area Command schedule and timeline.
f. Resolve conflicts between incident "realities" and agency executive "wants."
g. Establish appropriate location for the Area Command facilities.
h. Determine and implement an appropriate Area Command organization.
i. Determine need for Technical Specialists to support Area Command.
j. Obtain incident briefing and Incident Action Plans from Incident Commanders.

k. Assess incident situations prior to strategy meetings.
l. Conduct a joint meeting with all Incident Commanders.
m. Review objectives and strategies for each incident.
n. Periodically review critical resource needs.
o. Maintain a close coordination with the agency executive.
p. Establish priorities for use of critical resources.
q. Review procedures for interaction within the Area Command.
r. Approve Incident Commanders' requests for and release of critical resources.
s. Coordinate and approve Demobilization Plans.
t. Maintain log of major actions/decisions.

ASSISTANT AREA COMMANDER, PLANNING - The ACPC is responsible for collecting information from incident management teams in order to assess and evaluate potential conflicts in establishing incident objectives, strategies and the priority use of critical resources:

a. Obtain briefing from Area Commander.
b. Assemble information on individual incident objectives and begin to identify potential conflicts and/or ways for incidents to develop compatible operations.
c. Recommend the priorities for allocation of critical resources to incidents.
d. Maintain status on critical resource totals (not detailed status).
e. Ensure that advance planning beyond the next operational period is being accomplished.
f. Prepare and distribute Area Commander's decisions or orders.
g. Prepare recommendations for the reassignment of critical resources as they become available.
h. Ensure Demobilization Plans are coordinated between incident management teams and agency dispatchers.

i.   Schedule strategy meeting with Incident Commanders to conform to their planning processes.
j.   Prepare Area Command briefings as requested or needed.
k.   Maintain log of major actions/decisions.

ASSISTANT AREA COMMANDER, LOGISTICS - The ACLC is responsible for providing facilities, services and material at the Area Command level, and for ensuring effective use of critical resources and supplies among the incident management teams:

a.   Obtain briefing from the Area Commander.
b.   Provide facilities, services and materials for the Area Command organization.
c.   In the absence of the Area Commander Aviation Coordinator, ensure coordinated airspace temporary flight restrictions are in place and understood.
d.   Ensure coordinated communication links and frequencies are in place.
e.   Assist in the preparation of Area Command decisions.
f.   Ensure the continued effective and priority use of critical resources among the incident management teams.
g.   Maintain log of major actions/decisions.

AREA COMMAND AVIATION COORDINATOR - Technical Specialist responsible for ensuring effective use of critical aviation resources among multiple management teams:

a.   Obtains briefing from Area Commander.
b.   Coordinates with local unit(s) aviation managers, dispatch centers, and aviation facility managers.
c.   Monitors incident(s) aviation cost, efficiency, and safety. Ensures agency rules, regulations, and safety procedures are followed.

d. Provide to incidents local initial attack forces and other interested parties with an area aviation plan that outlines Area Command aviation procedures and specifics of the area aviation operation.

e. Allocates air and ground based aviation resources according to Area Command priorities and objectives.

f. Ensures inter-incident movement of aircraft is planned and coordinated.

g. Coordinates with local and adjacent initial attack aircraft bases and local dispatch to ensure that procedures for transiting incident area and corridors are in place. Ensures flight following procedures, entry/exit routes and corridors, hazards, frequencies and incident air space are known to all affected.

h. Coordinates with Incident Air Operations Branch Directors, dispatch, FAA, DOD, and local aviation authorities and administrators to ensure that Temporary Flight Restrictions are in place, coordinated, and do not overlap. Ensures that potential risks of operating on, near, or within Military Training Routes and Special-Use Airspace have been mitigated.

i. Ensures that a process is in place for timely transmittal of incident reports and oversees the process to ensure corrective action is taken.

j. Coordinates with incident, dispatch, and coordination centers to determine availability and status of committed and uncommitted of aviation resources, and to give status reports and situation appraisals for aviation assets and resources.

k. Coordinate with Incident Air Operations Branch Directors, Communications Unit Leaders, frequency coordinators, coordination centers and initial attack dispatch to establish coordinated aviation communications plans to ensure aviation communications plans to ensure aviation frequency management.

l. Coordinates and manages aviation program and operations if aviation assets are assigned to Area Command.

m. Coordinates the scheduling and movement of aviation safety assistance teams among incidents.

n. Assists incidents by coordinating with Contracting Officers, local aviation managers, and vendors concerning a variety of issues (fueling, contract modifications, contract extensions, etc.).

o. Coordinates with military officials and agency representatives concerning the assignments, utilization, status, and disposition of military aviation assets.

CHAPTER 4

COMPLEX

A complex is two or more individual incidents located in the same general proximity assigned to a single Incident Commander or Unified Command to facilitate management. These incidents are typically limited in scope and complexity and can be managed by a single entity.

The diagrams below illustrate a number of incidents in the same general proximity. These incidents may be identified as Branches or Divisions within the Operations Section.

Management responsibility for all of these incidents has been assigned to a single incident management team. A single incident may be complex, but it is not referred to as a "Complex." A complex may be in place with or without the use of Unified and/or Area Command.

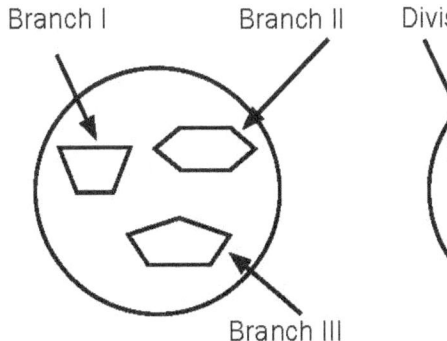
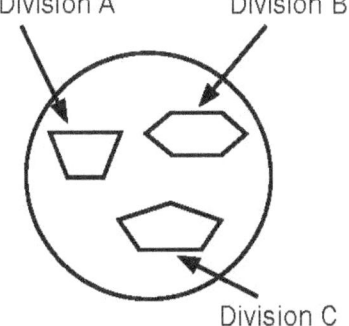

A typical organization would be as follows:

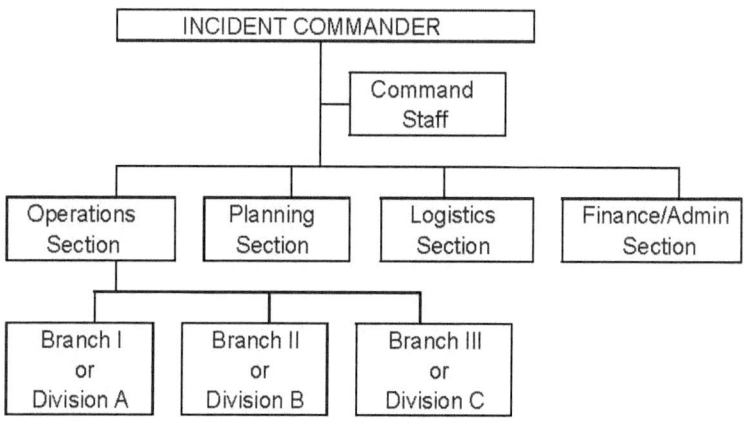

CHAPTER 5

COMMAND

Contents 5-1

ORGANIZATION CHART

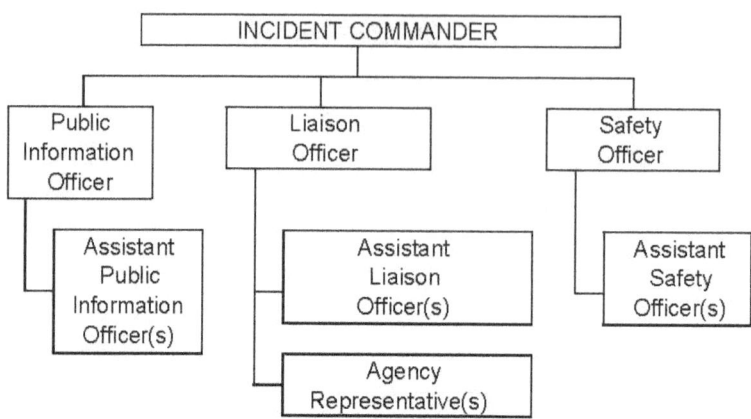

ESTABLISHMENT AND TRANSFER OF COMMAND

The highest-ranking official of the jurisdictional agency (ies) at the scene of the incident initially establishes Command. The Incident Commander is responsible for overall management of the incident. It is his/her responsibility to prepare the Incident Objectives that, in turn, will be the foundation upon which subsequent incident action planning will be based. Incident Objectives will be based on the requirements of the agency and the incident. They should be broad, measurable and follow an ordered sequence of events.

The Transfer of Command checklist below provides a basic guideline that can be used in almost any incident situation. This information may be captured on the Incident Briefing (ICS Form 201). However, agency policies and incident specific issues may require alterations to the transfer of command process.

When it is determined that a Transfer of Command (face-to-face) briefing needs to take place, the minimum essential information should include the following:

a. Situation Status
b. Objectives and Priorities
c. Current Organization
d. Resource Assignments
e. Resources En Route and/or Ordered
f. Facilities Established
g. Communications Plan
h. Prognosis, Concerns – Related Issues

As incidents grow in size or complexity, most agencies will transfer command one or more times. Whenever the transfer of command briefing takes place, the information conveyed should be recorded and displayed for easy retrieval and subsequent briefings.

POSITION CHECKLISTS

INCIDENT COMMANDER - The IC's responsibility is the overall management of the incident. On most incidents, a single IC carries out the command activity. However, Unified Command may be appropriate. The IC is selected by qualifications and experience.

The Incident Commander may have a Deputy, who may be from the same agency, or from an assisting agency. Deputies may also be used at section and branch levels of the ICS organization. Deputies must have the same qualifications as the person for whom they work for, as they must be ready to take over that position at any time:

a. Review Common Responsibilities (Page 1-2).
b. Assess the situation and/or obtain a briefing from the prior Incident Commander.
c. Determine Incident Objectives and strategy.
d. Establish the immediate priorities.
e. Establish an Incident Command Post.
f. Consider the need for Unified Command.
g. Establish an appropriate organization.
h. Ensure planning meetings are scheduled as required.
i. Approve and authorize the implementation of an Incident Action Plan.
j. Ensure that adequate safety and personnel accountability measures are in place.
k. Coordinate activity for all Command and General Staff.
l. Coordinate with key people and officials.
m. Approve requests for additional resources or for the release of resources.
n. Keep agency administrator informed of incident status.
o. Approve the use of trainees, volunteers, and auxiliary personnel.
p. Authorize release of information to the news media.
q. Ensure Incident Status Summary (ICS Form 209) is completed and forwarded to appropriate higher authority.
r. Order the demobilization of the incident when appropriate.
s. Maintain Unit/Activity Log (ICS Form 214).

Delegation of Authority: A statement provided to the Incident Commander by the Agency Executive delegating authority and assigning responsibility. The Delegation of Authority can include objectives, priorities, expectations, constraints, and other considerations or guidelines as needed. Many agencies require written Delegation of Authority to be given to Incident Commanders prior to their assuming command on larger incidents.

PUBLIC INFORMATION OFFICER - The PIO is responsible for developing and releasing information about the incident to the news media, to incident personnel, and to other appropriate agencies and organizations.

Only one Public Information Officer will be assigned for each incident, including incidents operating under Unified Command and multi-jurisdiction incidents. The Public Information Officer may have Assistant Public Information Officers as necessary, and the Assistant Public Information Officers may also represent assisting agencies or jurisdictions.

Agencies have different policies and procedures relative to the handling of public information. The following are the major responsibilities of the Public Information Officer that would generally apply on any incident:

a. Review Common Responsibilities (Page 1-2).
b. Determine from the Incident Commander if there are any limits on information release.
c. Develop material for use in media briefings.
d. Obtain Incident Commander's approval of media releases.
e. Coordinate with Joint Information Center (JIC) if established.
f. Inform media and conduct media briefings.
g. Arrange for tours and other interviews or briefings that may be required.
h. Obtain media information that may be useful to incident planning.
i. Maintain current information summaries and/or displays on the incident and provide information on status of incident to assigned personnel.
j. Assign Assistant Public Information Officers as appropriate.
k. Maintain Unit/Activity Log (ICS Form 214).

LIAISON OFFICER - Incidents that are multi-jurisdictional, or have several agencies involved, may require the establishment of the LOFR position on the Command Staff.

Only one Liaison Officer will be assigned for each incident, including incidents operating under Unified Command and multi-jurisdiction incidents. The Liaison Officer may have assistants as necessary, and the assistants may also represent assisting agencies or jurisdictions. The Liaison Officer is the point of contact for the Agency Representatives assigned to the incident by assisting or cooperating agencies.

a. Review Common Responsibilities (Page 1-2).
b. Be a contact point for Agency Representatives.
c. Maintain a list of assisting and cooperating agencies and Agency Representatives.
d. Assist in establishing and coordinating interagency contacts.
e. Keep agencies supporting the incident aware of incident status.
f. Monitor incident operations to identify current or potential inter-organizational problems.
g. Participate in planning meetings, providing current resource status, including limitations and capability of assisting agency resources.
h. Assign Assistant Liaison Officer(s) as appropriate.
i. Maintain Unit/Activity Log (ICS Form 214).

AGENCY REPRESENTATIVES - In many multi-jurisdiction incidents, an agency or jurisdiction will send a representative to assist in coordination efforts.

An Agency Representative is an individual assigned to an incident from an assisting or cooperating agency who has been delegated authority to make decisions on matters affecting that agency's participation at the incident.

Agency Representatives report to the Liaison Officer or to the Incident Commander in the absence of a Liaison Officer:

a. Review Common Responsibilities (Page 1-2).
b. Ensure that all agency resources are properly checked in at the incident.
c. Obtain briefing from the Liaison Officer or Incident Commander.
d. Inform assisting or cooperating agency personnel on the incident that the Agency Representative position for that agency has been filled.
e. Attend briefings and planning meetings as required.
f. Provide input on the use of agency resources unless resource technical specialists are assigned from the agency.
g. Cooperate fully with the Incident Commander and the General Staff on agency involvement at the incident.
h. Ensure the well being of agency personnel assigned to the incident.
i. Advise the Liaison Officer of any special agency needs or requirements.
j. Report to home agency dispatch or headquarters on a prearranged schedule.
k. Ensure that all agency personnel and equipment are properly accounted for and released prior to departure.
l. Ensure that all required agency forms, reports and documents are complete prior to departure.
m. Have a debriefing session with the Liaison Officer or Incident Commander prior to departure.
n. Maintain Unit/Activity Log (ICS Form 214).

SAFETY OFFICER - The SOF's function is to develop and recommend measures for assuring personnel safety, and to assess and/or anticipate hazardous and unsafe situations. Having full authority of the Incident Commander, the SOF can exercise emergency authority to stop or prevent unsafe acts.

Only one Safety Officer will be assigned for each incident. The Safety Officer may have Assistant Safety Officers as necessary, and the Assistant Safety Officers may also come from assisting agencies or jurisdictions as appropriate. Assistant Safety Officers may have specific responsibilities such as air operations, urban search and rescue, hazardous materials, or for specific geographic or functional areas of the incident:

a. Review Common Responsibilities (Page 1-2).
b. Participate in planning meetings, and advocate effective risk management.
c. Identify hazardous situations associated with the incident.
d. Review the Incident Action Plan for safety implications.
e. Exercise emergency authority to stop or prevent unsafe acts and communicate such exercise of authority to the Incident Command.
f. Investigate accidents that have occurred within the incident area.
g. Assign Assistant Safety Officers as needed.
h. Conduct and prepare an Incident Safety Analysis (ICS Form 215-AG/AW) as appropriate.
i. Initiate appropriate mitigation measures, i.e., Personnel Accountability, Fireline EMT's, Rapid Intervention Crew/ Company, etc.
j. Develop and communicate an incident safety message as appropriate.
k. Review and approve the Medical Plan (ICS Form 206).

l. Review and approve the Site Safety and Control Plan (ICS Form 208) as required.

m. Maintain Unit/Activity Log (ICS Form 214).

COMMAND AND GENERAL STAFF PLANNING CYCLE GUIDE

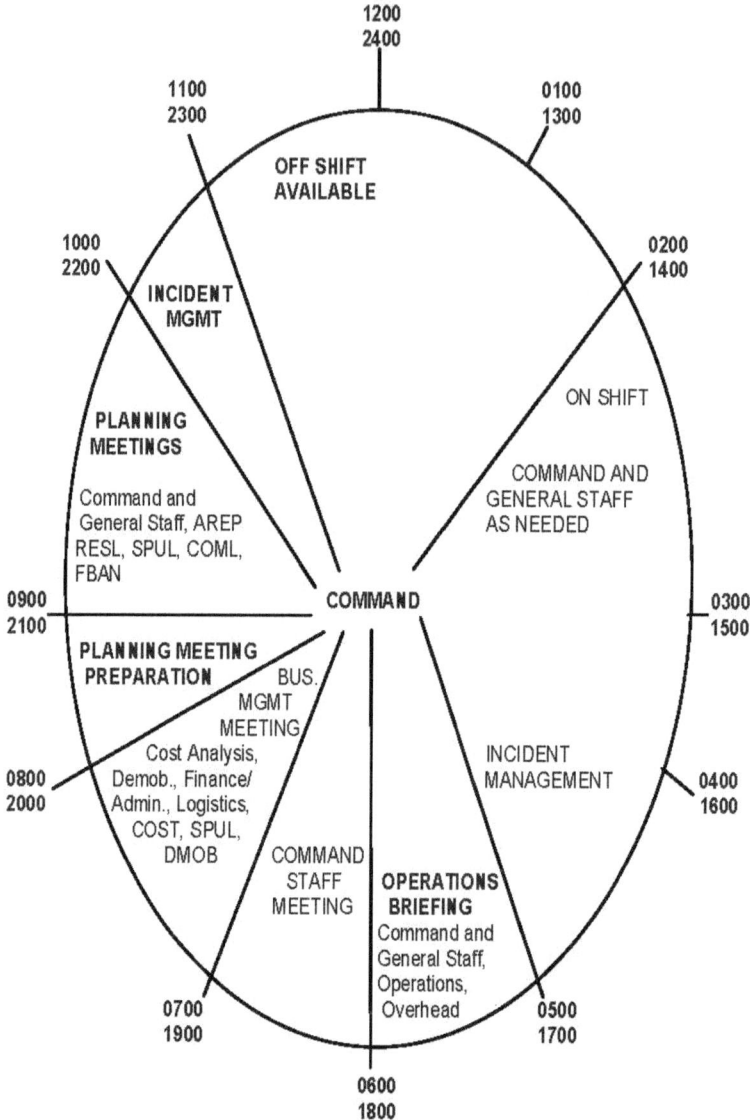

Example Based on 12-Hour Operational Period

CHAPTER 6

UNIFIED COMMAND

Contents 6-1

UNIFIED COMMAND

Experience has proven that at incidents involving multi-agencies, there is a critical need for integrating management of resources into one operational organization that is managed and supported by one command structure. This is best established through an integrated, multi-disciplined organization. In the ICS, employing what is known as Unified Command fills this critical need.

Unified Command is a team effort that allows all agencies with jurisdictional responsibility for an incident, either geographical or functional, to participate in the management of the incident. Developing and implementing a common set of incident objectives and strategies demonstrate this participation that all can subscribe to, without losing or abdicating agency authority, responsibility or accountability. Those organizations that participate in Unified Command should have statutory responsibility for some portion of the incident or event. Assisting and cooperating agencies with no statutory responsibility that nonetheless contribute resources to the incident should not function at the Unified Command level. These agencies should instead, assign Agency Representative to effectively represent their agencies and resources through the Liaison Officer. In these ways, the principles that define Unified Command provide all of the necessary mechanisms for organizational representation and interagency management within a multi-agency incident response.

At a local level, frequent training and realistic exercises involving those agencies that may be represented at actual incidents should be considered a prerequisite for successful management of multi-agency incidents. These activities serve to familiarize each participating agency of their respective roles and responsibilities and clarify the capabilities and limitations of each agency. For example, a planned event such as a parade or air show may provide an opportunity for local, state and federal agencies to operate in a Unified Command structure.

A successfully managed multi-agency incident will occur only when the participating agencies' personnel have confidence in each other's competencies, authorities, responsibilities, and limitations as they relate to the incident. Beyond the associated processes, guidelines, and exercises, is the requirement for an attitude of cooperation. Coordinated strategy, tactics, and resource utilization to accomplish incident control must be the focus of all agencies at the scene.

Within a Unified Command, one person is selected as spokesperson for the groups. Typically, the person representing the agency with the highest resource commitment or most visible activity on the incident is selected. In some cases, this task may simply be assigned to the person with the most experience.

Unified Command incorporates the following principles:

a. One set of objectives is developed for the entire incident.
b. A collective approach to developing strategies to achieve incident goals.
c. Improved information flow and coordination between all jurisdictions and agencies involved in the incident.
d. All agencies with responsibility for the incident have an understanding of one another's priorities and restrictions.
e. No agency's authority or legal requirements will be compromised or neglected.
f. Each agency is fully aware of the plans, actions and constraints of all others.

g. The combined efforts of all agencies are optimized as they perform their respective assignments under a single Incident Action Plan.

h. Duplicative efforts are reduced or eliminated, thus reducing cost and chances for frustration and conflict.

INITIAL UNIFIED COMMAND MEETING CHECKLIST

It is essential to begin unified planning as early as possible. Initiate Unified Command as soon as two or more agencies having jurisdictional or functional responsibilities come together on an incident. It is especially important on those incidents where there may be competing priorities based on agency responsibilities.

All of the jurisdictional agency's Incident Commanders need to get together before the first operational period planning meeting in an Initial Unified Command Meeting. This meeting provides the responsible agency officials with an opportunity to discuss and concur on important issues prior to joint incident action planning. The agenda for the command meeting should include the following:

a. State jurisdictional/agency priorities and objectives.
b. Present jurisdictional limitations, concerns, and restrictions.
c. Develop a collective set of incident objectives.
d. Establish and agree on acceptable priorities.
e. Adopt an overall strategy or strategies to accomplish objectives.
f. Agree on the basic organization structure.
g. Designate the most qualified and acceptable Operations Section Chief.
h. The Operations Section Chief will normally be from the jurisdiction or agency that has the greatest involvement in the incident, although that is not essential.

i. Agree on General Staff personnel designations and planning, logistical, and finance agreements and procedures.
j. Agree on the resource ordering process to be followed.
k. Agree on cost-sharing procedures.
l. Agree on informational matters.
m. Designate one agency official to act as the Unified Command spokesperson.

The members of the Unified Command must be authorized to perform certain activities and actions on behalf of the jurisdiction or agency they represent. Such activities include, ordering of additional resources in support of the Incident Action Plan, possible loaning or sharing of resources to other jurisdictions, and agree to financial cost-sharing arrangements with participating agencies.

COMMAND MEETING REQUIREMENTS

Unified Incident Commanders should meet prior to the Incident Planning Meeting to discuss a number of key items. This meeting will serve to clarify issues and provide direction to other incident personnel who will develop the formal Incident Action Plan.

The following checklist provides a series of items to be addressed during the meeting between Incident Commanders where the development of incident strategy and objectives is done:

a. The Command Meeting should include only agency Incident Commanders.
b. The meeting should be brief, and important points should be documented. The important points should include agency capabilities and limitations, functional and jurisdictional responsibilities and the individual agency's objectives.

c.   Prior to the meeting, the respective responsible officials
     should have reviewed the purposes and agenda items
     described above, and are prepared to discuss them.

The end result of the planning process will be a single
Incident Action Plan that addresses multi-jurisdiction or multi-
agency priorities and objectives, and provides an appropriate
level of tactical direction and resource assignments for the
unified effort.

CHAPTER 7

PLANNING PROCESS

Contents

PLANNING PROCESS

The checklist below provides basic steps appropriate for use in almost any incident situation. However, not all incidents require written plans and the need for written plans and attachments is based on incident requirements and the decision of the Incident Commander.

The Planning Checklist is to be used with the Operational Planning Worksheet (ICS Form 215-G/W). For more detailed instructions, see Planning Section Chief Position Manual (ICS 221-1). The Operations Section Chief should have a draft Operational Planning Worksheet (ICS Form 215-G/W) and the Safety Officer should have a draft Incident Safety Analysis (ICS Form 215-AG/AW) completed prior to the planning meeting.

Incident Objectives and strategy should be established before the planning meeting. For this purpose it may be necessary to hold a strategy meeting prior to the planning meeting.

The Planning Process works best when the incident is divided into logical geographical and/or functional units. The tactics and resources are then determined for each of the planning units and then the planning units are combined into divisions/groups utilizing span-of-control guidelines.

The ICS Form 215-G/W (Operational Planning Worksheet Generic and Wildland) and the ICS Form 215-AG/AW (Incident Safety Analysis – Generic and Wildland) are used to support the incident's planning process. They provide the Incident Commander, Command and General Staff with the means to identify Division or Group assignments, identify available and needed resources, and address safety considerations. During this process, safety issues identified must be mitigated or new tactics developed which adequately address safety concerns.

CHECKLIST   **PRIMARY RESPONSIBILITY**

1. Briefing on situation and resource status .... PSC
2. Set/review incident objectives IC
3. Plot control lines, establish Branch and Division
   boundaries, identify Group assignments .... OSC
4. Specify tactics for each Division/Group .... OSC
5. Specify safety mitigation measures for
   identified hazards in Divisions/Groups .... SOF
6. Specify resources needed by Division/Group ...... OSC, PSC
7. Specify Operations facilities and reporting
   locations – Plot on map .... OSC, PSC, LSC
8. Develop resource and personnel order .... LSC
9. Consider Communications, Medical, and
   Traffic Plan requirements .... PSC, LSC
10. Finalize, approve and implement Incident
    Action Plan .... PSC, IC, OSC

     IC   = Incident Commander
     PSC = Planning Section Chief
     OSC = Operations Section Chief
     LSC = Logistics Section Chief
     SOF = Safety Officer

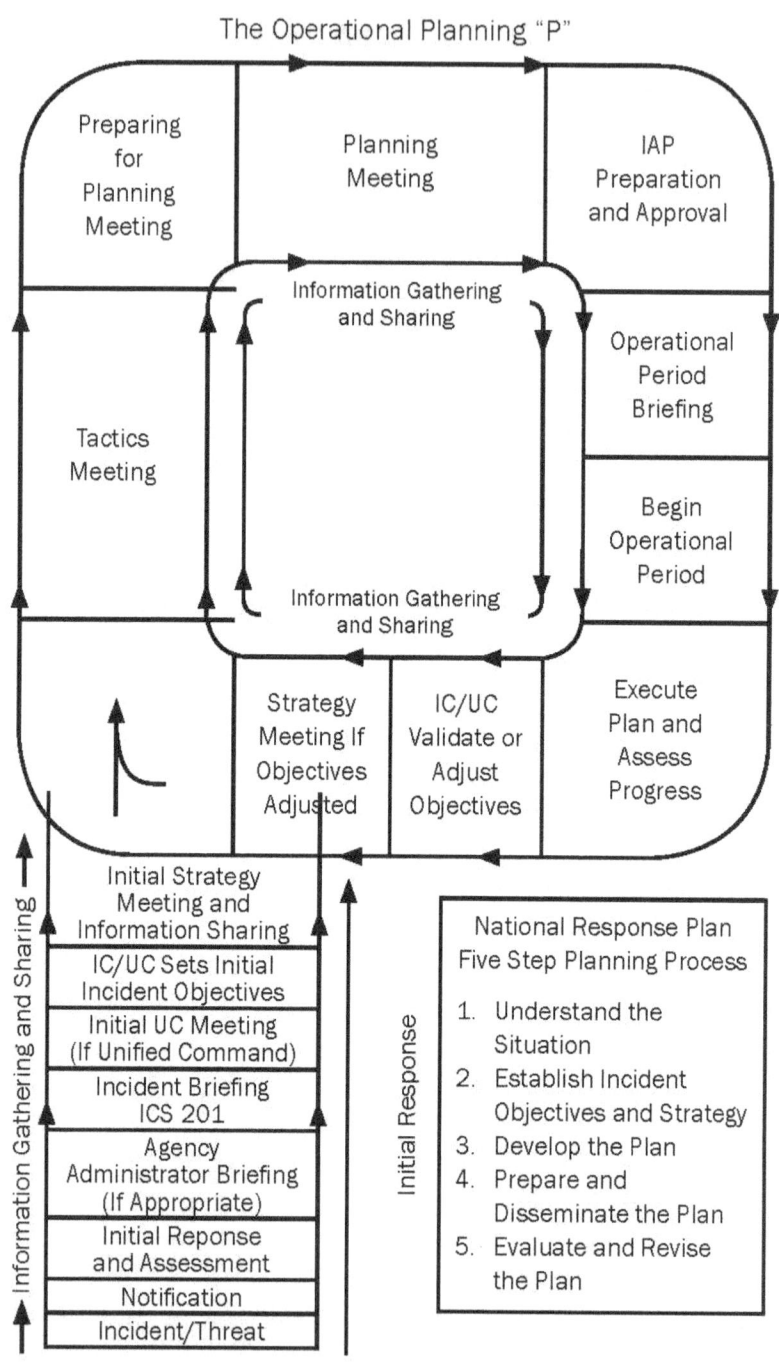

The Operational Planning "P"

Preparing for Planning Meeting	Planning Meeting	IAP Preparation and Approval

Information Gathering and Sharing

Tactics Meeting

Operational Period Briefing

Begin Operational Period

Information Gathering and Sharing

| Strategy Meeting If Objectives Adjusted | IC/UC Validate or Adjust Objectives | Execute Plan and Assess Progress |

Initial Strategy Meeting and Information Sharing

IC/UC Sets Initial Incident Objectives

Initial UC Meeting (If Unified Command)

Incident Briefing ICS 201

Agency Administrator Briefing (If Appropriate)

Initial Reponse and Assessment

Notification

Incident/Threat

Information Gathering and Sharing

Initial Response

National Response Plan Five Step Planning Process

1. Understand the Situation
2. Establish Incident Objectives and Strategy
3. Develop the Plan
4. Prepare and Disseminate the Plan
5. Evaluate and Revise the Plan

CHAPTER 8

OPERATIONS SECTION

Contents 8-1

ORGANIZATION CHART

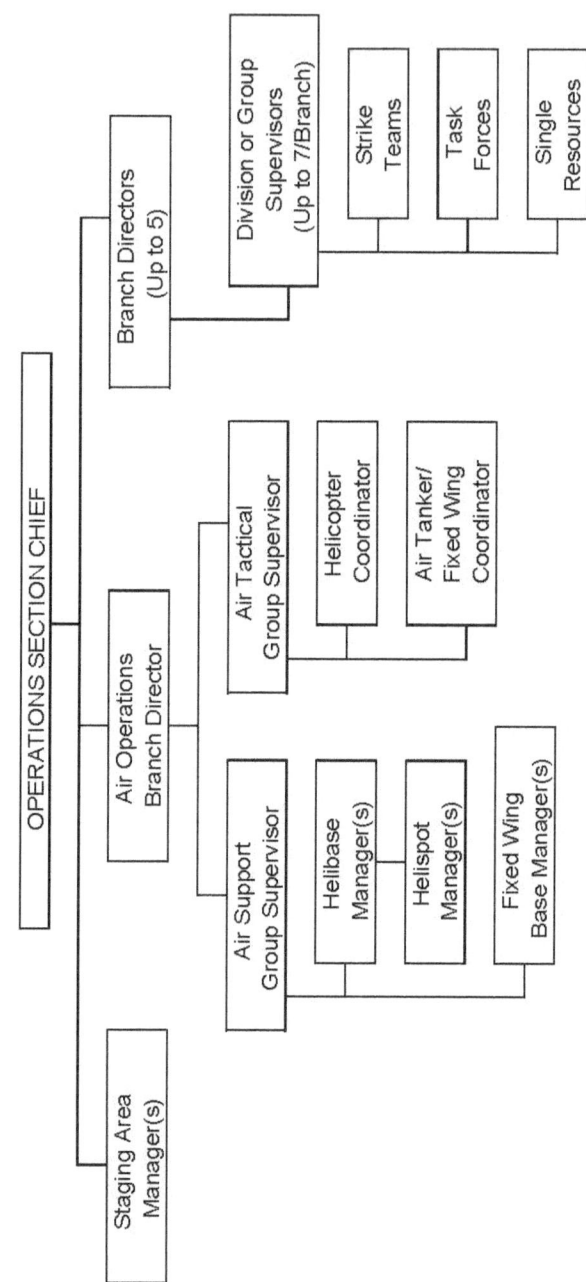

OPERATIONS SECTION CHIEF

Staging Area Manager(s)

Air Operations Branch Director

Branch Directors (Up to 5)

Air Support Group Supervisor

Air Tactical Group Supervisor

Division or Group Supervisors (Up to 7/Branch)

Helibase Manager(s)

Helispot Manager(s)

Fixed Wing Base Manager(s)

Helicopter Coordinator

Air Tanker/ Fixed Wing Coordinator

Strike Teams

Task Forces

Single Resources

POSITION CHECKLISTS

OPERATIONS SECTION CHIEF - The OSC, a member of the General Staff, is responsible for the management of all operations directly applicable to the primary mission ensuring the overall safety and welfare of all Section personnel. The OSC activates and supervises organization elements in accordance with the Incident Action Plan and directs its execution. The OSC also directs the preparation of unit operational plans, requests or releases resources, makes expedient changes to the Incident Action Plan as necessary, and reports such to the Incident Commander. The Deputy Operations Section Chief may be assigned for specific tasks, i.e., planning operations, day/night operations, evacuation or contingency planning, etc.:

a. Review Common Responsibilities (Page 1-2).
b. Develop the operations portion of the Incident Action Plan and complete the appropriate ICS Form 215 (G/W) as appropriate.
c. Brief and assign Operations Section personnel in accordance with Incident Action Plan.
d. Supervise Operations Section ensuring safety and welfare of all personnel.
e. Determine need and request additional resources.
f. Review suggested list of resources to be released and initiate recommendation for release of resources.
g. Assemble and disassemble Strike Teams and Task Forces assigned to Operations Section.
h. Report information about special activities, events, and occurrences to Incident Commander.
i. Maintain Unit/Activity Log (ICS Form 214).

OPERATIONS BRANCH DIRECTOR – OPBD's are under the direction of the Operations Section Chief, and are responsible for the implementation of the portion of the Incident Action Plan appropriate to the geographical and functional Branches:

a. Review Common Responsibilities (Page 1-2).
b. Develop with subordinates, alternatives for Branch control operations.
c. Attend planning meetings at the request of the Operations Section Chief.
d. Review Division/Group Assignment Lists (ICS Form 204) for Divisions or Groups within Branch. Modify lists based on effectiveness of current operations.
e. Assign specific work tasks to Division and Group Supervisors.
f. Supervise Branch operations.
g. Resolve logistical problems reported by subordinates.
h. Report to the Operations Section Chief when the Incident Action Plan needs to be modified, or additional resources are needed, or surplus resources are available, or when hazardous situations or significant events occur.
i. Approve accident and medical reports (home agency forms) originating within the Branch.
j. Maintain Unit/Activity Log (ICS Form 214).

DIVISION OR GROUP SUPERVISOR – DIVS's report to the Operations Chief (or Operations Branch Director when activated). The Supervisor is responsible for the implementation of the assigned portion of the Incident Action Plan. They are also responsible for the assignment of resources within the Division or Group, reporting on the progress of control operations, and the status of resources within the Division or Group. Division Supervisors are assigned to a specific geographical area of an incident. Group Supervisors are assigned to accomplish specific functions within the incident (i.e. Hazardous Material, Medical):

a. Review Common Responsibilities (Page 1-2).
b. Implement Incident Action Plan for Division or Group.
c. Provide Incident Action Plan to Strike Team Leaders, when available.
d. Identify increments assigned to the Division or Group.
e. Review assignments and incident activities with subordinates and assign tasks.
f. Ensure that Incident Communications and/or Resources Unit are advised of all changes in status of resources assigned to the Division or Group.
g. Coordinate activities with adjacent Divisions or Groups.
h. Determine need for assistance on assigned tasks.
i. Submit situation and resources status information to Operations Branch Directors or Operations Section Chief.
j. Report hazardous situations, special occurrences, or significant events (e.g., accidents, sickness) to immediate supervisor.
k. Ensure that assigned personnel and equipment get to and from assignments in a timely and orderly manner.
l. Resolve logistics problems within the Division or Group.
m. Participate in the development of tactical plans for next operational period.
n. Maintain Unit/Activity Log (ICS Form 214).

STRIKE TEAM or TASK FORCE LEADER - The Strike Team Leader or Task Force Leader reports to a Division Supervisor or Group Supervisor and is responsible for performing tactical assignments assigned to the Strike Team or Task Force. The Leader reports work progress and status of resources, maintains work records on assigned personnel, and relays other important information to their supervisor:

a. Review Common Responsibilities (Page 1-2).
b. Review assignments with subordinates and assign tasks.
c. Monitor work progress and make changes when necessary.
d. Coordinate activities with adjacent strike teams, task forces and single resources.
e. Travel to and from active assignment area with assigned resources.
f. Retain control of assigned resources while in available or out-of-service status.
g. Submit situation and resource status information to Division/Group Supervisor.
h. Maintain Unit/Activity Log (ICS Form 214).

STRUCTURE PROTECTION SPECIALIST – The STPS is a technical advisor to the Operations Section Chief or the Planning Section Chief. The recommendations of the STPS will be based on the incident objectives outlined in the IAP and identify the major components required to complete a Structure Protection Plan for threatened structures due to wildfire. The STPS will organize and implement this plan utilizing the recommended resources:

a. Review Common Responsibilities (Page 1-2).
b. Obtain reporting criteria and briefing from Operations Section Chief or Planning Section Chief.
c. Identify structure threat based on expected fire behavior.
d. Identify needed components to prepare Structure Protection Plan.
e. Develop LCES Plan related to structure protection.
f. Identify resource needs to carry out the plan.
g. Coordinate with local law enforcement agencies to carry out evacuation plan.
h. Brief all resources assigned to Branch, Division or Groups.
i. Ensure personnel safety.
j. Maintain Unit/Activity Log (ICS Form 214).

SINGLE RESOURCE - The person in charge of a single tactical resource will carry the unit designation of the resource:

a. Review Common Responsibilities (Page 1-2).
b. Review assignments.
c. Obtain necessary equipment/supplies.
d. Review weather/environmental conditions for assignment area.
e. Brief subordinates on safety measures.
f. Monitor work progress.
g. Ensure adequate communications with supervisor and subordinates.
h. Keep supervisor informed of progress and any changes.
i. Inform supervisor of problems with assigned resources.
j. Brief relief personnel, and advise them of any change in conditions.
k. Return equipment and supplies to appropriate unit.
l. Complete and turn in all time and use records on personnel and equipment.
m. Maintain Unit/Activity Log (ICS Form 214).

STAGING AREA MANAGER - The STAM is responsible for managing all activities within a Staging Area:

a. Review Common Responsibilities (Page 1-2).
b. Proceed to Staging Area.
c. Establish Staging Area layout.
d. Determine any support needs for equipment, feeding, sanitation and security.
e. Establish check-in function as appropriate.
f. Post areas for identification and traffic control.
g. Request maintenance service for equipment at Staging Area as appropriate.
h. Respond to request for resource assignments. (Note: This may be direct from Operations Section or via the Incident Communications Center).

i. Obtain and issue receipts for radio equipment and other supplies distributed and received at Staging Area.
j. Determine required resource levels from the Operations Section Chief.
k. Advise the Operations Section Chief when reserve levels reach minimums.
l. Maintain and provide status to Resources Unit of all resources in Staging Area.
m. Maintain Staging Area in orderly condition.
n. Demobilize Staging Area in accordance with Incident Demobilization Plan.
o. Maintain Unit/Activity Log (ICS Form 214).

AIR OPERATIONS BRANCH DIRECTOR - The AOBD, who is ground based, is primarily responsible for preparing the air operations portion of the Incident Action Plan. The plan will reflect agency restrictions that have an impact on the operational capability or utilization of resources (e.g., night flying, hours per pilot). After the plan is approved, Air Operations is responsible for implementing its strategic aspects--those that relate to the overall incident strategy as opposed to those that pertain to tactical operations (specific target selection).

Additionally, the Air Operations Branch Director is responsible for providing logistical support to helicopters operating on the incident. The Air Tactical Group Supervisor working with ground and air resources normally performs specific tactical activities (such as target selection and suggested modifications to specific tactical actions in the Incident Action Plan):

a. Review Common Responsibilities (Page 1-2).
b. Organize preliminary air operations.
c. Request declaration (or cancellation) of restricted air space area, (FAA Regulation 91.137).

d. Participate in preparation of the Incident Action Plan through Operations Section Chief. Insure that the Air Operations portion of the Incident Action Plan takes into consideration the Air Traffic Control requirements of assigned aircraft.
e. Perform operational planning for air operations.
f. Prepare and provide Air Operations Summary (ICS Form 220) to the Air Support Group and Fixed-Wing Bases.
g. Determine coordination procedures for use by air organization with ground Branches, Divisions or Groups.
h. Coordinate with appropriate Operations Section personnel.
i. Supervise all Air Operations activities associated with the incident.
j. Evaluate Helibase locations.
k. Establish procedures for emergency reassignment of aircraft.
l. Schedule approved flights of non-incident aircraft in the restricted air space area.
m. Coordinate and schedule infrared aircraft flights.
n. Coordinate with Operations Coordination Center (OCC) through normal channels on incident air operations activities.
o. Inform the Air Tactical Group Supervisor of the air traffic situation external to the incident.
p. Consider requests for non-tactical use of incident aircraft.
q. Resolve conflicts concerning non-incident aircraft.
r. Coordinate with Federal Aviation Administration (FAA).
s. Update air operations plans.
t. Report to the Operations Section Chief on air operations activities.
u. Report special incidents/accidents.
v. Arrange for an accident investigation team when warranted.
w. Maintain Unit/Activity Log (ICS Form 214).

AIR TACTICAL GROUP SUPERVISOR - The ATGS is primarily responsible for the coordination of aircraft operations when fixed and/or rotary-wing aircraft are operating on an incident. The ATGS performs these coordination activities while airborne. The ATGS reports to the Air Operations Branch Director:

a. Review Common Responsibilities (Page 1-2).
b. Determine what aircraft (air tankers and helicopters) are operating within area of assignment.
c. Manage air tactical activities based upon Incident Action Plan.
d. Establish and maintain communications and Air Traffic Control with pilots, Air Operations, Helicopter Coordinator, Air Tanker/Fixed Wing Coordinator, Air Support Group (usually Helibase Manager), and fixed wing support bases.
e. Coordinate approved flights of non-incident aircraft or non-tactical flights in restricted air space area.
f. Obtain information about air traffic external to the incident.
g. Receive reports of non-incident aircraft violating restricted air space area.
h. Make tactical recommendations to approved ground contact (Operations Section Chief, Operations Branch Director, or Division/Group Supervisor).
i. Inform Air Operations Branch Director of tactical recommendations affecting the air operations portion of the Incident Action Plan.
j. Report on Air Operations activities to the Air Operations Branch Director. Advise Air Operations immediately if aircraft mission assignments are causing conflicts in the Air Traffic Control System.
k. Report on incidents/accidents.
l. Maintain Unit/Activity Log (ICS Form 214).

HELICOPTER COORDINATOR - The HLCO is primarily responsible for coordinating tactical or logistical helicopter mission(s) at the incident. The HLCO can be airborne or on the ground operating from a high vantage point. The HLCO reports to the Air Tactical Group Supervisor. Activation of this position is contingent upon the complexity of the incident and the number of helicopters assigned. There may be more than one HLCO assigned to an incident:

a. Review Common Responsibilities (Page 1-2).
b. Determine what aircraft (air tankers and helicopters) are operating within incident area of assignment.
c. Survey assigned incident area to determine situation, aircraft hazards and other potential problems.
d. Coordinate Air Traffic Control with pilots, Air Operations Branch Director, Air Tactical Group Supervisor, Air Tanker/ Fixed Wing Coordinator and the Air Support Group (usually Helibase Manager) as the situation dictates.
e. Coordinate the use of assigned ground-to-air and air-to-air communications frequencies with the Air Tactical Group Supervisor, Communications Unit, or local agency dispatch center.
f. Ensure that all assigned helicopters know appropriate operating frequencies.
g. Coordinate geographical areas for helicopter operations with Air Tactical Group Supervisor and make assignments.
h. Determine and implement air safety requirements and procedures.
i. Ensure that approved night flying procedures are in operation.
j. Receive assignments, brief pilots, assign missions, and supervise helicopter activities.
k. Coordinate activities with Air Tactical Group Supervisor, Air Tanker/Fixed Wing Coordinator, Air Support Group and ground personnel.

l.   Maintain continuous observation of assigned helicopter-operating area and inform Air Tactical Group Supervisor of incident conditions including any aircraft malfunction or maintenance difficulties, and anything that may affect the incident.

m.   Inform Air Tactical Group Supervisor when mission is completed and reassign helicopter as directed.

n.   Request assistance or equipment as required.

o.   Report incidents or accidents to Air Operations Branch Director and Air Tactical Group Supervisor immediately.

p.   Maintain Unit/Activity Log (ICS Form 214).

AIR TANKER/FIXED WING COORDINATOR - The ATCO is primarily responsible for coordinating assigned air tanker operations at the incident. The Coordinator, who is always airborne, reports to the Air Tactical Group Supervisor. Activation of this position is contingent upon the need or upon complexity of the incident:

a.   Review Common Responsibilities (Page 1-2).

b.   Determine all aircraft including air tankers and helicopters operating within incident area of assignment.

c.   Survey incident area to determine situation, aircraft hazards and other potential problems.

d.   Coordinate the use of assigned ground-to-air and air-to-air communication frequencies with Air Tactical Group Supervisor, Communications Unit or local agency dispatch center and establish air tanker air-to-air radio frequencies.

e.   Ensure air tankers know appropriate operating frequencies.

f.   Determine incident air tanker capabilities and limitations for specific assignments.

g.   Coordinate Air Traffic Control with pilots, Air Operations Branch Director, Air Tactical Group Supervisor, Helicopter Coordinator, and Air Support Group (usually Helibase Manager) as the situation dictates.

h.   Determine and implement air safety requirement procedures.

i.   Receive assignments, brief pilots, assign missions, and supervise fixed-wing activities.

j.   Coordinate activities with Air Tactical Group Supervisor, Helicopter Coordinator and ground operations personnel.

k.   Maintain continuous observation of air tanker operating areas.

l.   Provide information to ground resources, if necessary.

m.   Inform Air Tactical Group Supervisor of overall incident conditions including aircraft malfunction or maintenance difficulties.

n.   Inform Air Tactical Group Supervisor when mission is completed and reassign air tankers as directed.

o.   Request assistance or equipment as necessary.

p.   Report incidents or accidents immediately to Air Operations Branch Director.

q.   Maintain Unit/Activity Log (ICS Form 214).

AIR SUPPORT GROUP SUPERVISOR - The ASGS is primarily responsible for supporting and managing Helibase and Helispot operations and maintaining liaison with fixed-wing air bases. This includes providing: 1) fuel and other supplies, 2) maintenance and repair of helicopters, 3) retardant mixing and loading, 4) keeping records of helicopter activity, and 5) providing enforcement of safety regulations. These major functions are performed at Helibases and Helispots. Helicopters during landing and take-off and while on the ground are under the control of the Air Support Group's Helibase Manager or Helispot Manager. The ASGS reports to the Air Operations Branch Director:

a. Review Common Responsibilities (Page 1-2).
b. Obtain copy of the Incident Action Plan from the Air Operations Branch Director including Air Operations Summary (ICS Form 220).
c. Participate in Air Operations Branch Director planning activities.
d. Inform Air Operations Branch Director of group activities.
e. Identify resources/supplies dispatched for Air Support Group.
f. Request special air support items from appropriate sources through Logistics Section.
g. Identify Helibase and Helispot locations (from Incident Action Plan) or from Air Operations Branch Director.
h. Determine need for assignment of personnel and equipment at each Helibase and Helispot.
i. Coordinate special requests for air logistics.
j. Maintain coordination with airbases supporting the incident.
k. Coordinate activities with Air Operations Branch Director.
l. Obtain assigned ground-to-air frequency for Helibase operations from Communications Unit Leader or Incident Radio Communications Plan (ICS Form 205).
m. Inform Air Operations Branch Director of capability to provide night-flying service.
n. Ensure compliance with each agency's operations checklist for day and night operations.
o. Ensure dust abatement procedures are implemented at Helibase and Helispots.
p. Provide aircraft rescue firefighting service for Helibases and Helispots.
q. Ensure that Air Traffic Control procedures are established between Helibase and Helispots and the Air Tactical Group Supervisor, Helicopter Coordinator or Air Tanker/ Fixed Wing Coordinator.
r. Maintain Unit/Activity Log (ICS Form 214).

HELIBASE MANAGER - The HEB2/1 has primary responsibility for managing all activities at the assigned Helibase:

a. Review Common Responsibilities (Page 1-2).
b. Obtain Incident Action Plan including Air Operations Summary (ICS Form 220).
c. Participate in Air Support Group planning activities.
d. Inform Air Support Supervisor of Helibase activities.
e. Report to assigned Helibase. Brief pilots and other assigned personnel.
f. Manage resources/supplies dispatched to Helibase.
g. Ensure Helibase is posted and cordoned.
h. Coordinate Helibase Air Traffic control with pilots, Air Support Group Supervisor, Air Tactical Group Supervisor, Helicopter Coordinator and the Takeoff and Landing Controller.
i. Manage retardant mixing and loading operations.
j. Ensure helicopter fueling, maintenance and repair services are provided.
k. Supervise manifesting and loading of personnel and cargo.
l. Ensure dust abatement techniques are provided and used at Helibases and Helispots.
m. Ensure security is provided at each Helibase and Helispot.
n. Ensure aircraft rescue firefighting services are provided for the Helibase.
o. Request special air support items from the Air Support Group Supervisor.
p. Receive and respond to special requests for air logistics.
q. Supervise personnel responsible to maintain agency records, reports of helicopter activities, and Check-In List (ICS Form 211).
r. Coordinate activities with Air Support Group Supervisor.

s. Display organization and work schedule at each Helibase, including Helispot organization and assigned radio frequencies.
t. Solicit pilot input concerning selection and adequacy of Helispots, communications, Air Traffic Control, operational difficulties, and safety problems.
u. Maintain Unit/Activity Log (ICS Form 214).

HELISPOT MANAGER – The HESM is supervised by the Helibase Manager and is responsible for providing safe and efficient management of all activities at the assigned Helispot:

a. Review Common Responsibilities (Page 1-2).
b. Obtain Incident Action Plan including Air Operations Summary (ICS Form 220).
c. Report to assigned Helispot.
d. Coordinate activities with Helibase Manager.
e. Inform Helibase Manager of Helispot activities.
f. Manage resources/supplies dispatch to Helispot.
g. Request special air support items from Helibase Manager.
h. Coordinate Air Traffic Control and Communications with pilots, Helibase Manager, Helicopter Coordinator, Air Tanker/Fixed Wing Coordinator and Air Tactical Group Supervisor when appropriate.
i. Ensure aircraft rescue firefighting services are available.
j. Ensure that dust control is adequate, debris cannot blow into rotor system, touchdown zone slope is not excessive and rotor clearance is sufficient.
k. Supervise or perform retardant loading at Helispot.
l. Perform manifesting and loading of personnel and cargo.
m. Coordinate with pilots for proper loading and unloading and safety problems.
n. Maintain agency records and reports of helicopter activities.
o. Maintain Unit/Activity Log (ICS Form 214).

MIXMASTER - The MXMS is responsible for providing fire retardant to helicopters at the rate specified and for the expected duration of job. The MXMS reports to the Helibase Manager:

a.   Review Common Responsibilities (Page 1-2).
b.   Obtain Air Operations Summary (ICS Form 220).
c.   Check accessory equipment, such as valves, hoses and storage tanks.
d.   Take immediate steps to get any items and personnel to do the job.
e.   Plan the specific layout to conduct operations.
f.   Determine if water or retardant is to be used and which helicopters may have load restrictions.
g.   Maintain communication with Helibase Manager.
h.   Supervise the crew in setting up operations.
i.   Supervise crew in loading retardant into helicopters.
j.   Make sure supply of retardants is kept ahead of demand.
k.   Attend to the safety and welfare of crew.
l.   See that the base is cleaned up before leaving.
m.   Keep necessary agency records.
n.   Maintain Unit/Activity Log (ICS Form 214).

DECK COORDINATOR - The DECK is responsible for providing coordination of a Helibase landing area for personnel and cargo movement. The DECK reports to the Helibase Manager:

a.   Review Common Responsibilities (Page 1-2).
b.   Obtain Air Operations Summary (ICS Form 220).
c.   Establish emergency landing areas.
d.   Ensure deck personnel understand aircraft rescue firefighting procedures.
e.   Establish and mark landing pads.

f.   Ensure sufficient personnel are available to load and unload personnel and cargo safely.
g.   Ensure deck area is properly posted.
h.   Provide for vehicle control.
i.   Supervise deck management personnel (Loadmasters and Parking Tenders).
j.   Ensure dust abatement measures are met.
k.   Ensure that all assigned personnel are posted to the daily organization chart.
l.   Ensure proper manifesting and load calculations are done.
m.   Ensure Air Traffic Control operation is coordinated with Takeoff and Landing Coordinator.
n.   Maintain agency records.
o.   Maintain Unit/Activity Log (ICS Form 214).

LOADMASTER (PERSONNEL/CARGO) - The LOAD is responsible for the safe operation of loading and unloading of cargo and personnel at a Helibase. The LOAD reports to the Deck Coordinator:

a.   Review Common Responsibilities (Page 1-2).
b.   Obtain Air Operations Summary (ICS Form 220).
c.   Ensure proper posting of loading and unloading areas.
d.   Perform manifesting and loading of personnel and cargo.
e.   Ensure sling load equipment is safe.
f.   Know aircraft rescue firefighting procedures.
g.   Supervise loading and unloading crews.
h.   Coordinate with Take Off and Landing Coordinator.
i.   Maintain Unit/Activity Log (ICS Form 214).

PARKING TENDER - The PARK is responsible for the takeoff and landing of helicopters at an assigned helicopter pad. The PARK reports to the DECK. A PARK should be assigned for each helicopter pad:

a. Review Common Responsibilities (Page 1-2).
b. Supervise activities at the landing pad (personnel and helicopter movement, vehicle traffic, etc.).
c. Know and understand the aircraft rescue firefighting procedures.
d. Ensure agency checklist is followed.
e. Ensure helicopter pilot needs are met at the landing pad.
f. Ensure landing pad is properly maintained (dust abatement, marking, etc.).
g. Ensure landing pad is properly marked.
h. Check personnel seatbelts, cargo restraints and helicopter doors.
i. Maintain Unit/Activity Log (ICS Form 214).

TAKEOFF AND LANDING COORDINATOR - The TOLC is responsible for providing coordination of arriving and departing helicopters at a Helibase and all helicopter movement on and around the Helibase. The TOLC reports to the Helibase Manager:

a. Review Common Responsibilities (Page 1-2).
b. Obtain Air Operations Summary (ICS Form 220).
c. Check radio system before commencing operation.
d. Coordinate with radio operation on helicopter flight routes and patterns.
e. Maintain communications with all incoming and outgoing helicopters.
f. Maintain constant communications with radio operator.
g. Coordinate with Deck Coordinator and Parking Tender before commencing operation and during operation.
h. Maintain Unit/Activity Log (ICS Form 214).

HELICOPTER TIMEKEEPER - The HETM is responsible for keeping time on all helicopters assigned to the Helibase. HETM reports to the Radio Operator:

a.   Review Common Responsibilities (Page 1-2).
b.   Obtain Air Operations Summary (ICS Form 220).
c.   Determine number of helicopters by agency.
d.   Determine helicopter time needed by agency.
e.   Record operation time of helicopters.
f.   Fill out necessary agency time reports.
g.   Obtain necessary timekeeping forms.
h.   Maintain Unit/Activity Log (ICS Form 214).

OPERATIONS SECTION PLANNING CYCLE GUIDE

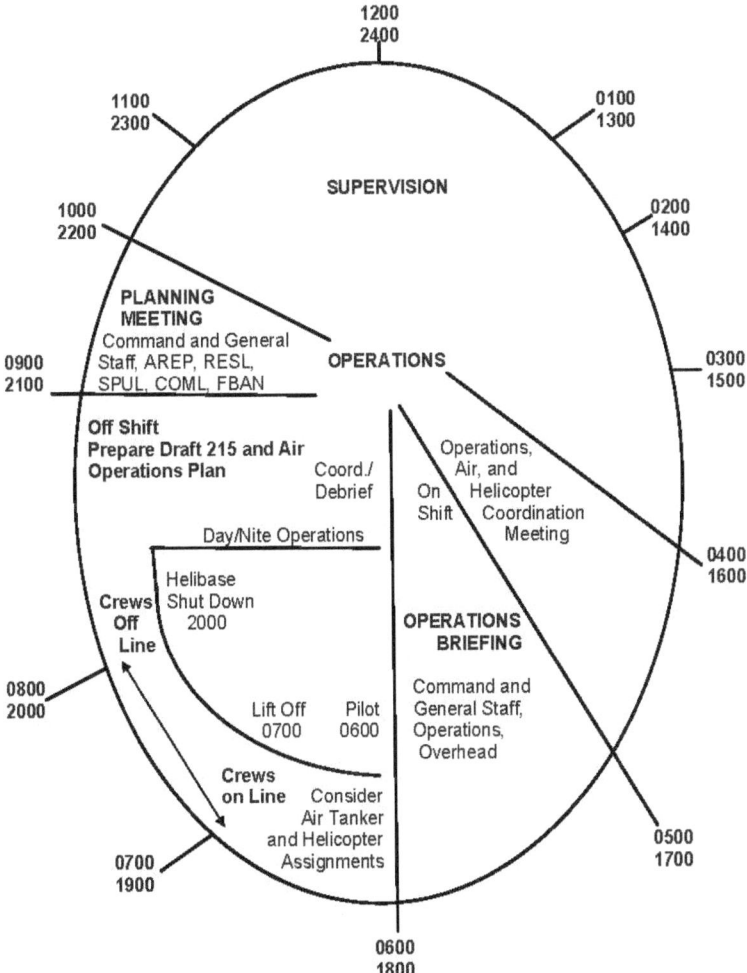

Example Based on 12-Hour Operational Period

Notes

CHAPTER 9

PLANNING SECTION

Contents 9-1

ORGANIZATION CHART

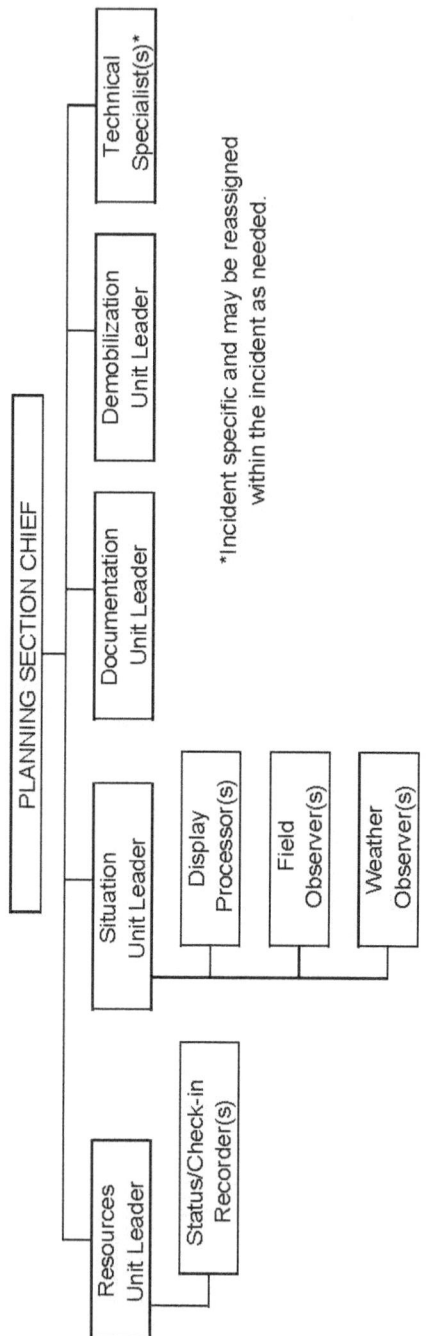

PLANNING SECTION CHIEF

Resources Unit Leader
- Status/Check-in Recorder(s)

Situation Unit Leader
- Display Processor(s)
- Field Observer(s)
- Weather Observer(s)

Documentation Unit Leader

Demobilization Unit Leader

Technical Specialist(s)*

*Incident specific and may be reassigned within the incident as needed.

PLANNING

PLANNING

POSITION CHECKLISTS

PLANNING SECTION CHIEF - The PSC, a member of the Incident Commander's General Staff, is responsible for the collection, evaluation, dissemination and use of information about the development of the incident and status of resources. The PSC is responsible for facilitating the Planning Process as described in Chapter 7. The PSC is also responsible for ensuring the safety and welfare of all Section personnel. Information is needed to: 1) understand the current situation, 2) predict probable course of incident events, and 3) prepare alternative strategies and control operations for the incident:

a. Review Common Responsibilities (Page 1-2).
b. Collect and process situation information about the incident.
c. Supervise preparation of the Incident Action Plan.
d. Provide input to the Incident Commander and Operations Section Chief in preparing the Incident Action Plan.
e. Reassign out-of-service personnel already on-site to ICS organizational positions as appropriate.
f. Establish information requirements and reporting schedules for Planning Section Units (e.g., Resources Unit and Situation Unit).
g. Determine need for any specialized resources in support of the incident.
h. If requested, assemble and disassemble strike teams and task forces not assigned to Operations.
i. Establish special information collection activities as necessary, e.g., weather, environmental, toxics, etc.
j. Assemble information on alternative strategies.
k. Provide periodic predictions on incident potential.
l. Report any significant changes in incident status.
m. Compile and display incident status information.

n. Oversee preparation and implementation of Incident Demobilization Plan.
o. Incorporate plans, (e.g., Traffic, Medical, Communications, Site Safety) into the Incident Action Plan.
p. Maintain Unit/Activity Log (ICS Form 214).

RESOURCES UNIT LEADER - The RESL is responsible for maintaining the status of all assigned resources (primary and support) at an incident. This is achieved by overseeing the check-in of all resources, maintaining a status-keeping system indicating current location and status of all resources, and maintenance of a master list of all resources, e.g., key supervisory personnel, primary and support resources, etc.:

a. Review Common Responsibilities (Page 1-2).
b. Review Unit Leader Responsibilities (Page 1-2).
c. Establish check-in function at incident locations.
d. Prepare Organization Assignment List (ICS Form 203) and Organization Chart (ICS Form 207).
e. Prepare appropriate parts of Assignment Lists (ICS Form 204).
f. Prepare and maintain the Command Post display (to include organization chart and resource allocation and deployment).
g. Maintain and post the current status and location of all resources.
h. Maintain master roster of all resources checked in at the incident.
i. A Status/Check-In Recorder reports to the Resources Unit Leader and assists with the accounting of all incident-assigned resources.
j. Maintain Unit/Activity Log (ICS Form 214).

STATUS/CHECK-IN RECORDER – SCKN's are needed at each check-in location to ensure that all resources assigned to an incident are accounted for:

a. Review Common Responsibilities (Page 1-2).
b. Obtain required work materials, including Check-in Lists (ICS Form 211), Resource Status Cards (ICS Form 219), and status display boards.
c. Establish communications with the Communication Center and Ground Support Unit.
d. Post signs so that arriving resources can easily find incident check-in location(s).
e. Record check-in information on Check-in Lists (ICS Form 211).
f. Transmit check-in information to Resources Unit on regular prearranged schedule or as needed.
g. Forward completed Check-in Lists (ICS Form 211) to the Resources Unit.
h. Receive, record, and maintain resource status information on Resource Status Cards (ICS Form 219) for incident assigned Single Resources, Strike Teams, Task Forces, and Overhead personnel.
i. Maintain files of Check-in Lists (ICS Form 211).

SITUATION UNIT LEADER - The collection, processing and organizing of all incident information takes place within the Situation Unit. The Situation Unit may prepare future projections of incident growth, maps and intelligence information:

a. Review Common Responsibilities (Page 1-2).
b. Review Unit Leader Responsibilities (Page 1-3).
c. Begin collection and analysis of incident data as soon as possible.

d.   Prepare, post, or disseminate resource and situation status information as required, including special requests.
e.   Prepare periodic predictions or as requested.
f.   Prepare the Incident Status Summary (ICS Form 209).
g.   Provide photographic services and maps if required.
h.   Maintain Unit/Activity Log (ICS Form 214).

DISPLAY PROCESSOR - The DPRO is responsible for the display of incident status information obtained from Field Observers, resource status reports, aerial and orthography photographs and infrared data:

a.   Review Common Responsibilities (Page 1-2).
b.   Determine location of work assignment.
c.   Determine numbers, types and locations of displays required.
d.   Determine map requirements for Incident Action Plans.
e.   Determine time limits for completion.
f.   Obtain information from Situation Unit.
g.   Obtain necessary equipment and supplies.
h.   Obtain copy of Incident Action Plan for each operational period.
i.   Assist Situation Unit Leader in analyzing and evaluating field reports.
j.   Develop required displays in accordance with time limits for completion.
k.   Maintain Unit/Activity Log (ICS Form 214).

FIELD OBSERVER - The FOBS is responsible to collect situation information from personal observations at the incident and provide this information to the Situation Unit Leader:

a.   Review Common Responsibilities (Page 1-2).
b.   Obtain copy of Incident Action Plan for the Operational Period.

c. Obtain necessary equipment and supplies.
d. Identify all facility locations (e.g., Helispots, Division and Branch boundaries).
e. Report information to Situation Unit by established procedure.
f. Report immediately any condition observed that may cause danger and safety hazard to personnel.
g. Gather intelligence that will lead to accurate predictions.
h. Maintain Unit/Activity Log (ICS Form 214).

WEATHER OBSERVER - The WOBS is responsible to collect current incident weather information and provide the information to an assigned meteorologist, Fire Behavior Analyst or Situation Unit Leader:

a. Review Common Responsibilities (Page 1-2).
b. Obtain weather data collection equipment.
c. Obtain appropriate transportation to collection site(s).
d. Record and report weather observations at assigned locations on schedule.
e. Turn in equipment at completion of assignment.
f. Demobilize according to Incident Demobilization Plan.
g. Demobilize incident displays in accordance with Incident Demobilization Plan.
h. Maintain Unit/Activity Log (ICS Form 214).

DOCUMENTATION UNIT LEADER - The DOCL is responsible for the maintenance of accurate, up-to-date incident files. The Documentation Unit will also provide duplication services. Incident files will be stored for legal, analytical, and historical purposes:

a. Review Common Responsibilities (Page 1-2).
b. Review Unit Leader Responsibilities (Page 1-2).
c. Set up work area and begin organization of incident files.

d. Establish duplication service; respond to requests.
e. File all official forms and reports.
f. Review records for accuracy and completeness; inform appropriate units of errors or omissions.
g. Provide incident documentation as requested.
h. Store files for post-incident use.
i. Maintain Unit/Activity Log (ICS Form 214).

DEMOBILIZATION UNIT LEADER - The DMOB is responsible for developing the Incident Demobilization Plan. On large incidents, demobilization can be quite complex, requiring a separate planning activity. Note that not all agencies require specific demobilization instructions:

a. Review Common Responsibilities (Page 1-2).
b. Review Unit Leader Responsibilities (Page 1-2).
c. Review incident resource records to determine the likely size and extent of demobilization effort.
d. Based on above analysis, add additional personnel, workspace and supplies as needed.
e. Coordinate demobilization with Agency Representatives.
f. Monitor ongoing Operations Section resource needs.
g. Identify surplus resources and probable release time.
h. Develop incident checkout function for all units.
i. Evaluate logistics and transportation capabilities to support demobilization.
j. Establish communications with off-incident facilities, as necessary.
k. Develop an Incident Demobilization Plan detailing specific responsibilities and release priorities and procedures.
l. Prepare appropriate directories (e.g., maps, instructions, etc.) for inclusion in the Demobilization Plan.
m. Distribute Demobilization Plan (on and off-site).
n. Ensure that all Sections/Units understand their specific demobilization responsibilities.

o. Supervise execution of the Incident Demobilization Plan.

p. Brief Planning Section Chief on demobilization progress.

q. Maintain Unit/Activity Log (ICS Form 214).

TECHNICAL SPECIALISTS - Certain incidents or events may require the use of THSP who have specialized knowledge and expertise. THSP may function within the Planning Section, or be assigned wherever their services are required. Specific THSP have been identified (i.e. weather, fire behavior, etc.) and specific checklists are listed below or in the specific Operational System Description (i.e. US&R). For all other THSP not otherwise specified, use the checklist at the end of this section.

DAMAGE INSPECTION TECHNICAL SPECIALIST - The DINS is primarily responsible for inspecting damage and/or potential "at-risk" property, and natural resources. The DINS usually function within the Planning Section and may be assigned to the Situation Unit or can be reassigned wherever their services are required. Damage inspection includes loss of environmental resources, infrastructure, transportation, structures, and other real/personal property:

a. Review Common Responsibilities (Page 1-2).

b. Establish communications with local government representatives of effective jurisdictions.

c. Determine and order resources.

d. Determine coordination procedures with other sections, units and local agencies.

e. Establish work area, and obtain necessary supplies.

f. Collect information pertaining to incident causes losses.

g. Participate in Planning Section activities.

h. Prepare documentation as required.

i. Respond to requests for information from approved sources.

j.   Prepare final Situation Status Field Inspection Report (SSFIR), and forward to the Documentation Unit Leader.

k.   Maintain Unit/Activity Log (ICS Form 214).

ENVIRONMENTAL SPECIALIST – The ENSP is primarily responsible for accessing the potential impacts of an incident on the environment, determining environmental restrictions, recommending alternative strategies and priorities for addressing environmental concerns. The ENSP functions within the Planning Section as part of the Situation Unit:

a.   Review Common Responsibilities (Page 1-2).

b.   Participate in the development of the Incident Action Plan and review the general control objectives including alternative strategies.

c.   Collect and validate environmental information within the incident area by reviewing pre-attack land use and management plans.

d.   Determine environmental restrictions within the incident area.

e.   Develop suggested priorities for preservation of the environment.

f.   Provide environmental analysis information, as requested.

g.   Collect and transmit required records and logs to Documentation Unit at the end of each operational period.

h.   Maintain Unit/Activity Log (ICS Form 214).

FIRE BEHAVIOR ANALYST - The FBAN is primarily responsible for establishing a weather data collection system, and to develop required fire behavior predictions based on fire history, fuel, weather, and topography information:

a.   Review Common Responsibilities (Page 1-2).

b.   Establish weather data requirements.

c.   Verify dispatch of meteorologist.

d. Confirm that mobile weather station has arrived and is operational.
e. Inform meteorologist of weather data requirements.
f. Forward weather data to Planning Section Chief.
g. Collect, review and compile fire history data.
h. Collect, review and compile exposed fuel data.
i. Collect, review and compile information about topography and fire barriers.
j. Provide weather information and other pertinent information to Situation Unit Leader for inclusion in Incident Status Summary (ICS Form 209).
k. Review completed Incident Status Summary report and Incident Action Plan.
l. Prepare fire behavior prediction information at periodic intervals or upon request and forward to Planning Section Chief.
m. Maintain Unit/Activity Log (ICS Form 214).

GEOGRAPHICAL INFORMATION SYSTEM SPECIALIST - A GISS is responsible for spatial information collection, display, analysis, and dissemination. The GISS will provide Global Positioning System (GPS) support, integrate infrared data, and incorporate all relevant data to produce map products, statistical data for reports, and/or analyses. GISS usually functions within the Planning Section, or assigned wherever their services are required within the incident organization:

a. Review Common Responsibilities (Page 1-2).
b. Check in with the Status/Check-In Recorder.
c. Obtain briefing from appropriate supervisor.
d. Establish communication with local government representatives, of all affected jurisdictions, through the incident Liaison Officer.
e. Determine and order resources needed.
f. Determine coordination procedures with other sections, units, and local agencies.

g. Establish work area, and acquire work materials.
h. Obtain appropriate transportation and communications.
i. Determine the availability of needed GIS support products.
j. Participate in Planning Section activities.
k. Prepare GIS products as determined by supervisor.
l. Keep supervisor informed.
m. Respond to requests from approved sources for additional GIS products.
n. Prepare final GIS summary report consisting of all incident GIS products and forward to Documentation Unit Leader.
o. Maintain Unit/Activity Log (ICS Form 214).

RESOURCE USE SPECIALIST – The Resource Use Specialist is primarily responsible for advising incident personnel on the specific capabilities, limitations of certain specialized response resources. In addition, the Resource Specialist can recommend strategies for use of these resources:

a. Review Common Responsibilities (Page 1-2).
b. Participate in the development of the Incident Action Plan and review general control objectives including alternative strategies as requested.
c. Collect information on incident resources as needed.
d. Respond to requests for information about limitations and capabilities of resources.
e. Collect and transmit records and logs to Documentation Unit at the end of each operational period.
f. Maintain Unit/Activity Log (ICS Form 214).

TRAINING SPECIALIST – The TNSP coordinates incident training opportunities and activities, ensuring the quality of the training assignments and completing documentation of the incident training. The TNSP organizes and implements the incident-training program and analyzes and facilitates training assignments to fulfill individual development needs of trainees:

a. Review Common Responsibilities (Page 1-2).
b. Inform Planning Section Chief of planned use of trainees.
c. Review trainee assignments and modify if appropriate.
d. Coordinate the assignments of trainees to incident positions with Resources Unit.
e. Brief trainees and trainers on training assignments and objectives.
f. Coordinate use of unassigned trainees.
g. Make follow-up contacts on the job to provide assistance and advice for trainees to meet training objectives as appropriate and with approval of unit leaders.
h. Ensure trainees receive performance evaluation.
i. Monitor operational procedures and evaluate training needs.
j. Respond to requests for information concerning training activities.
k. Give Training Specialist records and logs to Documentation Unit at the end of each operational period.
l. Maintain Unit/Activity Log (ICS Form 214).

WATER RESOURCES SPECIALIST – The Water Resource Specialist is primarily responsible to advise incident personnel on the sources of fire suppression water, the capabilities of the water sources, and to assist in the development of additional systems or system capability to meet incident demands:

a. Review Common Responsibilities (Page 1-2).
b. Participate in the development of the Incident Action Plan and review general control objectives, including alternative strategies presently in effect.
c. Collect and validate water resource information within the incident area.

d. Prepare information on available water resources.

e. Establish water requirements needed to support fire suppression actions.

f. Compare Incident Control Objectives as stated in the Plan, with available water resources and report inadequacies or problems to Planning Section Chief.

g. Participate in the preparation of Incident Action Plan when requested.

h. Respond to requests for water information.

i. Collect and transmit records and logs to Documentation Unit at the end of each operational period.

j. Maintain Unit/Activity Log (ICS Form 214).

TECHNICAL SPECIALISTS (NOT OTHERWISE SPECIFIED):

a. Review Common Responsibilities (Page 1-2).

b. Check in with the Status/Check-In Recorder.

c. Obtain briefing from supervisor.

d. Obtain personal protective equipment as appropriate.

e. Determine coordination procedures with other sections, units, and local agencies.

f. Establish work area and acquire work materials.

g. Participate in the development of the Incident Action Plan and review the general control objectives including alternative strategies as appropriate.

h. Obtain appropriate transportation and communications.

i. Keep supervisor informed.

j. Maintain Unit/Activity Log (ICS Form 214).

PLANNING SECTION PLANNING CYCLE GUIDE

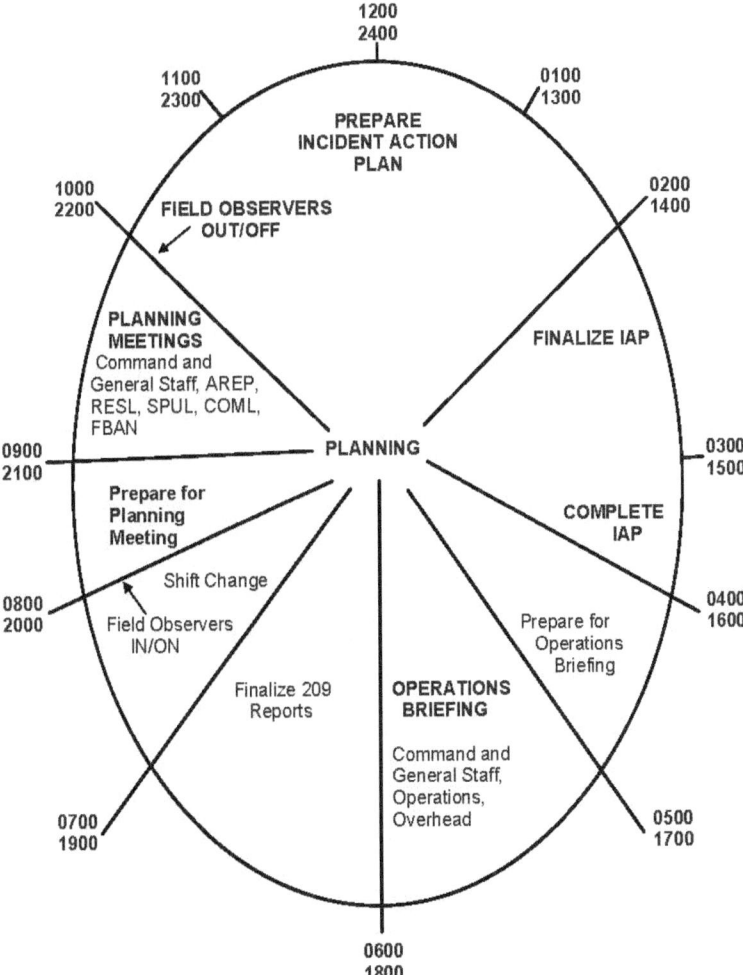

Example Based on 12-Hour Operational Period

Notes

CHAPTER 10

LOGISTICS SECTION

Contents 10-1

Organization Chart 10-2

Position Checklists 10-3

ORGANIZATION CHART

LOGISTICS SECTION CHIEF

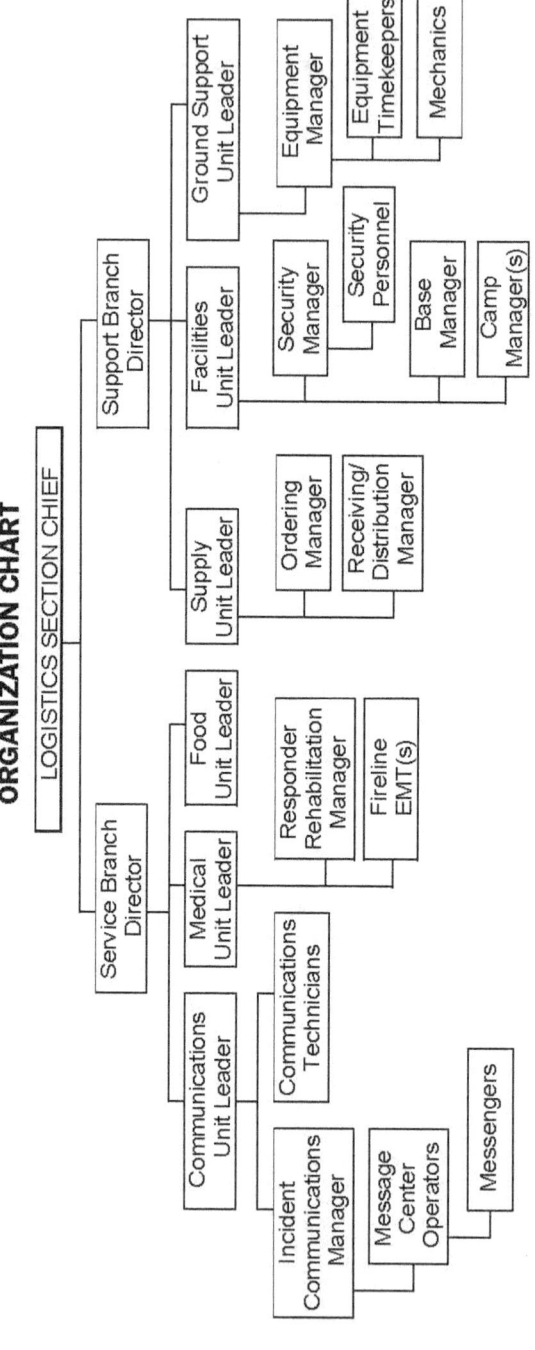

POSITION CHECKLISTS

LOGISTICS SECTION CHIEF - The LSC, a member of the General Staff, is responsible for providing facilities, services, and material in support of the incident. The LSC participates in development and implementation of the Incident Action Plan, activates and supervises assigned Branches/Units, and is responsible for the safety and welfare of Logistics Section personnel:

a. Review Common Responsibilities (Page 1-2).
b. Plan organization of Logistics Section.
c. Assign work locations and preliminary work tasks to Section personnel.
d. Notify Resources Unit of Logistics Section Units activated including names and locations of assigned personnel.
e. Assemble and brief Branch Directors and Unit Leaders.
f. Participate in preparation of Incident Action Plan.
g. Identify service and support requirements for planned and expected operations.
h. Provide input to and review Communications Plan, Medical Plan and Traffic Plan.
i. Coordinate and process requests for additional resources.
j. Review Incident Action Plan and estimate Section needs for next operational period.
k. Advise on current service and support capabilities.
l. Prepare service and support elements of the Incident Action Plan.
m. Estimate future service and support requirements.
n. Receive Demobilization Plan from Planning Section.
o. Recommend release of unit resources in conformity with Demobilization Plan.
p. Ensure general welfare and safety of Logistics Section personnel.
q. Maintain Unit/Activity Log (ICS Form 214).

SERVICE BRANCH DIRECTOR - The SVBD, when activated, is under the supervision of the Logistics Section Chief, and is responsible for the management of all service activities at the incident. The SVBD supervises the operations of the Communications, Medical and Food Units:

a.   Review Common Responsibilities (Page 1-2).
b.   Obtain working materials.
c.   Determine level of service required to support operations.
d.   Confirm dispatch of Branch personnel.
e.   Participate in planning meetings of Logistics Section personnel.
f.   Review Incident Action Plan.
g.   Organize and prepare assignments for Service Branch personnel.
h.   Coordinate activities of Branch Units.
i.   Inform Logistics Section Chief of Branch activities.
j.   Resolve Service Branch problems.
k.   Maintain Unit/Activity Log (ICS Form 214).

COMMUNICATIONS UNIT LEADER - The COML, under the direction of the Service Branch Director or Logistics Section Chief, is responsible for developing plans for the effective use of incident communications equipment and facilities; installing and testing of communications equipment; supervision of the Incident Communications Center; distribution of communications equipment to incident personnel; and the maintenance and repair of communications equipment:

a.   Review Common Responsibilities (Page 1-2).
b.   Review Unit Leader Responsibilities (Page 1-3).
c.   Determine unit personnel needs.
d.   Prepare and implement the Incident Radio Communications Plan (ICS Form 205).

e. Ensure the Incident Communications Center and Message Center are established.
f. Establish appropriate communications distribution/maintenance locations within Base/Camp(s).
g. Ensure communications systems components are installed, tested and maintained.
h. Ensure an equipment accountability system is established.
i. Ensure personal portable radio equipment from cache is distributed per Incident Radio Communications Plan (ICS Form 205).
j. Provide technical information as required.
k. Supervise Communications Unit activities.
l. Maintain records on all communications equipment as appropriate.
m. Ensure equipment is tested and repaired.
n. Recover equipment from relieved or released units.
o. Maintain Unit/Activity Log (ICS Form 214).

INCIDENT COMMUNICATIONS MANAGER - The INCM is responsible to receive and transmit radio and telephone messages among and between personnel and to provide dispatch services at the incident:

a. Review Common Responsibilities (Page 1-2).
b. Ensure adequate staffing (Incident Communications Manager).
c. Obtain and review Incident Action Plan to determine incident organization and Incident Radio Communications Plan (ICS Form 205).
d. Set up Incident Radio Communications Center - check out equipment.
e. Request service on any inoperable or marginal equipment.
f. Set up Message Center location as required.
g. Receive and transmit messages within and external to incident.

h. Maintain General Messages files.
i. Maintain a record of unusual incident occurrences.
j. Provide briefing to relief on current activities, equipment status, and any unusual communications situations.
k. Turn in appropriate documents to Incident Communications Manager or Communications Unit Leader.
l. Demobilize Communications Center in accordance with Incident Demobilization Plan.
m. Maintain Unit/Activity Log (ICS Form 214).

MEDICAL UNIT LEADER - The MEDL, under the direction of the Service Branch Director or Logistics Section Chief, is primarily responsible for the development of the Medical Plan (ICS Form 206), obtaining medical aid and transportation for injured and ill incident personnel, establishment of responder rehabilitation and preparation of reports and records:

a. Review Common Responsibilities (Page 1-2).
b. Review Unit Leader Responsibilities (Page 1-2).
c. Participate in Logistics Section/Service Branch planning activities.
d. Establish and staff Medical Unit.
e. Establish Responder Rehabilitation.
f. Prepare the Medical Plan (ICS Form 206).
g. Prepare procedures for major medical emergency.
h. Declare major medical emergency as appropriate.
i. Respond to requests for medical aid, medical transportation, and medical supplies.
j. Prepare and submit necessary documentation.
k. Maintain Unit/Activity Log (ICS Form 214).

RESPONDER REHABILITATION MANAGER – The Responder Rehabilitation Manager reports to the Medical Unit Leader and is responsible for the rehabilitation of incident personnel who are suffering from the effects of strenuous work and/or extreme conditions:

a. Review Common Responsibilities (Page 1-2).

b. Designate responder rehabilitation location and have location announced on radio with radio designation "Rehab."

c. Request necessary medical personnel to evaluate medical condition of personnel being rehabilitated.

d. Request necessary resources for rehabilitation of personnel, e.g., water, juice, personnel.

e. Request through Food Unit or Logistics Section Chief feeding as necessary for personnel being rehabilitated.

f. Release rehabilitated personnel to Operations Section or Planning Section for reassignment.

g. Maintain appropriate records and documentation.

h. Maintain Unit/Activity Log (ICS Form 214).

FIRELINE EMERGENCY MEDICAL TECHNICIAN – The FEMT provides emergency medical care to personnel operating on the fireline. The FEMT initially reports to the Medical Unit Leader, if established, or the Logistics Section Chief. The FEMT must establish and maintain liaison with, and respond to requests from the Operations Section personnel to whom they are subsequently assigned:

The checklist presented below should be considered as a minimum requirement for the position. Users of this manual may augment these lists as necessary. Note that some of the activities are one-time actions while others are ongoing for the duration of an incident:

a. Review Common Responsibilities (Page 1-2).

b. Check in and obtain briefing from the Logistics Section Chief, or the Medical Unit Leader if established. Briefing will include current incident situation, anticipated medical needs, and required local medical protocol including documentation.

c. Receive assignment and assess current situation.
d. Anticipate needs and obtain medical supplies from the incident.
e. Secure copies of local emergency medical service forms/paperwork if available.
f. Secure/check out portable radio with all incident frequencies.
g. Obtain a copy of the Incident Action Plan (IAP) and review the Medical Plan (ICS Form 206).
h. Identify and contact assigned tactical supervisor and confirm your travel route, transportation and ETA prior to leaving your check-in location.
i. Meet with assigned tactical supervisor and obtain briefing.
j. Obtain briefing from the FEMT you are relieving, if applicable
k. Upon arrival at your assigned location, perform a radio check with your assigned tactical supervisor, incident Communications Unit and the Medical Unit, if established.
l. Maintain ongoing contact and interaction with personnel on your assignment to assess medical needs and provide assistance when needed.
m. Make requests for transportation of ill and injured personnel, through channels, as outlined in the Medical Plan (ICS Form 206).
n. Make notifications of incident related illnesses and injuries as outlined in the Medical Plan (ICS Form 206).
o. At the conclusion of each shift, advise your tactical supervisor that you are departing and will report to the Medical Unit Leader for debriefing and submission of patient care documentation.
p. Secure operations and demobilize as outlined in the Demobilization Plan.

q. Maintain Unit/Activity Log (ICS Form 214).

FOOD UNIT LEADER – The FDUL is responsible for supplying the food needs for the entire incident, including all remote locations (e.g., Camps, Staging Areas), as well as providing food for personnel unable to leave tactical field assignments:

a. Review Common Responsibilities (Page 1-2).
b. Review Unit Leader Responsibilities (Page 1-3).
c. Determine food and water requirements.
d. Determine method of feeding to best fit each facility or situation.
e. Obtain necessary equipment and supplies and establish cooking facilities.
f. Ensure that well-balanced menus are provided.
g. Order sufficient food and potable water from the Supply Unit.
h. Maintain an inventory of food and water.
i. Maintain food service areas, ensuring that all appropriate health and safety measures are being followed.
j. Supervise caterers, cooks, and other Food Unit personnel as appropriate.
k. Maintain Unit/Activity Log (ICS Form 214).

SUPPORT BRANCH DIRECTOR – The SUBD, when activated, is under the direction of the Logistics Section Chief, and is responsible for development and implementation of logistics plans in support of the Incident Action Plan. The SUBD supervises the operations of the Supply, Facilities and Ground Support Units:

a. Review Common Responsibilities (Page 1-2).
b. Obtain work materials.

c. Identify Support Branch personnel dispatched to the incident.
d. Determine initial support operations in coordination with Logistics Section Chief and Support Branch Director.
e. Prepare initial organization and assignments for support operations.
f. Assemble and brief Support Branch personnel.
g. Determine if assigned Branch resources are sufficient.
h. Maintain surveillance of assigned units work progress and inform Logistics Section Chief of activities.
i. Resolve problems associated with requests from Operations Section.
j. Maintain Unit/Activity Log (ICS Form 214).

SUPPLY UNIT LEADER – The SPUL is primarily responsible for ordering personnel, equipment and supplies; receiving and storing all supplies for the incident; maintaining an inventory of supplies; and servicing non-expendable supplies and equipment:

a. Review Common Responsibilities (Page 1-2).
b. Review Unit Leader Responsibilities (Page 1-3).
c. Participate in Logistics Section/Support Branch planning activities.
d. Determine the type and amount of supplies en route.
e. Review Incident Action Plan for information on operations of the Supply Unit.
f. Develop and implement safety and security requirements.
g. Order, receive, distribute, and store supplies and equipment.
h. Receive and respond to requests for personnel, supplies and equipment.
i. Maintain inventory of supplies and equipment.
j. Service reusable equipment.
k. Submit reports to the Support Branch Director.

l.   Maintain Unit/Activity Log (ICS Form 214).

ORDERING MANAGER – The ODRM is responsible for placing all orders for supplies and equipment for the incident. The ODRM reports to the Supply Unit Leader:

a.   Review Common Responsibilities (Page 1-2).
b.   Obtain necessary agency (ies) order forms.
c.   Establish ordering procedures.
d.   Establish name and telephone numbers of agency personnel receiving orders.
e.   Set up filing system.
f.   Get names of incident personnel who have ordering authority.
g.   Check on what has already been ordered.
h.   Ensure order forms are filled out correctly.
i.   Place orders in a timely manner.
j.   Consolidate orders when possible.
k.   Identify times and locations for delivery of supplies and equipment.
l.   Keep Receiving and Distribution Manager informed of orders placed.
m.   Submit all ordering documents to Documentation Control Unit through Supply Unit Leader before demobilization.
n.   Maintain Unit/Activity Log (ICS Form 214).

RECEIVING AND DISTRIBUTION MANAGER – The RCDM is responsible for receiving and distribution of all supplies and equipment (other than primary resources) and the service and repair of tools and equipment. The RCDM reports to the Supply Unit Leader:

a.   Review Common Responsibilities (Page 1-2).
b.   Order required personnel to operate supply area.
c.   Organize physical layout of supply area.

d. Establish procedures for operating supply area.
e. Set up filing system for receiving and distribution of supplies and equipment.
f. Maintain inventory of supplies and equipment.
g. Develop security requirement for supply area.
h. Establish procedures for receiving supplies and equipment
i. Submit necessary reports to Supply Unit Leader.
j. Notify Ordering Manager of supplies and equipment received.
k. Provide necessary supply records to Supply Unit Leader.
l. Maintain Unit/Activity Log (ICS Form 214).

FACILITIES UNIT LEADER – The FACL is primarily responsible for the layout and activation of incident facilities, e.g., Base, Camp(s) and Incident Command Post. The Unit provides sleeping and sanitation facilities for incident personnel and manages Base and Camp(s) operations. Each facility (Base, Camp) is assigned a manager who reports to the FACL and is responsible for managing the operation of the facility. The basic functions or activities of the Base/Camp Manager are to provide security service, and general maintenance. The FACL reports to the Support Branch Director:

a. Review Common Responsibilities (Page 1-2).
b. Review Unit Leader Responsibilities (Page 1-3).
c. Receive a copy of the Incident Action Plan.
d. Participate in Logistics Section/Support Branch planning activities.
e. Determine requirements for each facility.
f. Prepare layouts of incident facilities.
g. Notify unit leaders of facility layout.
h. Activate incident facilities.
i. Provide Base/Camp Managers.
j. Provide sleeping facilities.
k. Provide security services.

l.   Provide facility maintenance services-sanitation, lighting, and cleanup.

m.   Maintain Unit/Activity Log (ICS Form 214).

FACILITY MAINTENANCE SPECIALIST – The FMNT is responsible to ensure that proper sleeping and sanitation facilities are maintained, provide shower facilities, maintain lights and other electrical equipment, and maintain the Base, Camp and Incident Command Post facilities in a clean and orderly manner:

a.   Review Common Responsibilities (Page 1-2).

b.   Request required maintenance support personnel and assign duties.

c.   Obtain supplies, tools, and equipment.

d.   Supervise/perform assigned work activities.

e.   Ensure that all facilities are maintained in a safe condition.

f.   Disassemble temporary facilities when no longer required.

g.   Restore area to pre-incident condition.

h.   Maintain Unit/Activity Log (ICS Form 214).

SECURITY MANAGER –The SECM is responsible to provide safeguards needed to protect personnel and property from loss or damage:

a.   Review Common Responsibilities (Page 1-2).

b.   Establish contacts with local law enforcement agencies as required.

c.   Contact the Resource Use Specialist for crews or Agency Representatives to discuss any special custodial requirements that may affect operations.

d.   Request required personnel support to accomplish work assignments.

e.   Ensure that support personnel are qualified to manage security problems.

f.   Develop Security Plan for incident facilities.
g.   Adjust Security Plan for personnel and equipment changes and releases.
h.   Coordinate security activities with appropriate incident personnel.
i.   Keep the peace, prevent assaults, and settle disputes through coordination with Agency Representatives.
j.   Prevent theft of all government and personal property
k.   Document all complaints and suspicious occurrences.
l.   Maintain Unit/Activity Log (ICS Form 214).

BASE/CAMP MANAGER – The BCMG is responsible to ensure that appropriate sanitation, security, and facility management services are conducted at all incident facilities.

On large incidents, a Base and one or more Camps may be established by the General Staff to provide better support to operations. Base is the location where the primary logistics functions are coordinated and administered. Camps are typically smaller in nature and more remote. Camps may be in place several days or may be moved depending upon the nature of the incident. Functional unit activities performed at the Base may be performed at the Camp(s). These activities could include, Supply Unit, Medical Unit, Ground Support Unit, Food Unit, Communications Unit, as well as the Facilities Unit functions of facility maintenance and security. Camp Managers are responsible to provide non-technical coordination for all units operating within the Camp. The General Staff will determine units assigned to Camps. Personnel requirements for units at Camps will be determined by the parent unit, based on kind and size of incident and expected duration of Camp operations. The Base/Camp Manager duties include:

a. Review Common Responsibilities (Page 1-2).
b. Determine personnel support requirements.
c. Obtain necessary equipment and supplies.
d. Ensure that all sanitation, shower and sleeping facilities are set up and properly functioning.
e. Make sleeping arrangements.
f. Provide direct supervision for all facility maintenance and security services at Base/Camp(s).
g. Ensure that strict compliance is made with all applicable safety regulations.
h. Ensure that all Base-to-Camp communications are centrally coordinated.
i. Ensure that all Base-to-Camp transportation scheduling is centrally coordinated.
j. Provide overall coordination of all Base/Camp activities to ensure that all assigned units operate effectively and cooperatively in meeting incident objectives.
k. Maintain Unit/Activity Log (ICS Form 214).

GROUND SUPPORT UNIT LEADER – The GSUL is primarily responsible for support of out-of-service resources; transportation of personnel, supplies, food, and equipment; fueling, service, maintenance, and repair of vehicles and other ground support equipment; and development and implementation of the Incident Traffic Plan:

a. Review Common Responsibilities (Page 1-2).
b. Review Unit Leader Responsibilities (Page 1-3).
c. Participate in Support Branch/Logistics Section planning activities.
d. Develop and implement Traffic Plan.
e. Support out-of-service resources.
f. Notify Resources Unit of all status changes on support and transportation vehicles.

g. Arrange for and activate fueling, maintenance, and repair of ground resources.
h. Maintain inventory of support and transportation vehicles (Support Vehicle Inventory ICS Form 218).
i. Provide transportation services.
j. Collect use information on rented equipment.
k. Requisition maintenance and repair supplies (e.g., fuel, spare parts).
l. Maintain incident roads.
m. Submit reports to Support Branch Director as directed.
n. Maintain Unit/Activity Log (ICS Form 214).

EQUIPMENT MANAGER – The EQPM provides service, repair and fuel for all apparatus and equipment; provides transportation and support vehicle services; and maintains records of equipment use and service provided:

a. Review Common Responsibilities (Page 1-2).
b. Obtain Incident Action Plan to determine locations for assigned resources, Staging Area locations, and fueling and service requirements for all resources.
c. Obtain necessary equipment and supplies.
d. Provide maintenance and fueling according to schedule.
e. Prepare schedules to maximize use of available transportation.
f. Provide transportation and support vehicles for incident use.
g. Coordinate with Agency Representatives on service and repair policies as required.
h. Inspect equipment condition and ensure coverage by equipment agreement.
i. Determine supplies (e.g., gasoline, diesel, oil and parts needed to maintain equipment in efficient operating condition), and place orders with Supply Unit.
j. Maintain Support Vehicle Inventory (ICS Form 218).

k. Maintain equipment rental records.

l. Maintain equipment service and use records.

m. Check all service repair areas to ensure that all appropriate safety measures are being taken.

n. Maintain Unit/Activity Log (ICS Form 214).

LOGISTICS SECTION PLANNING CYCLE GUIDE

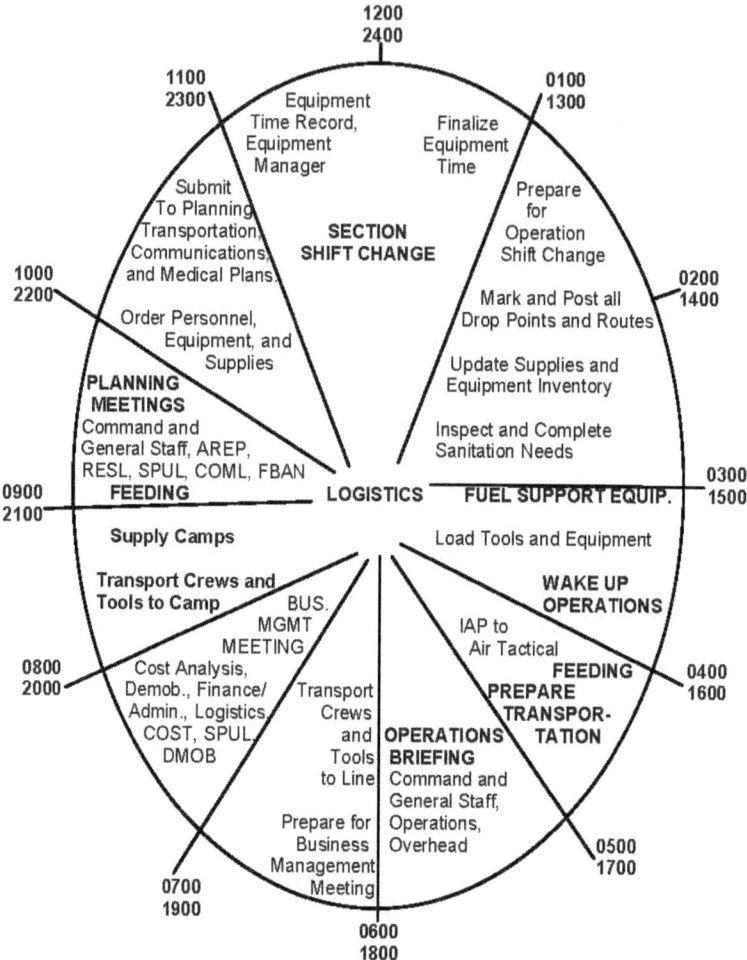

Example Based on 12-Hour Operational Period

CHAPTER 11

FINANCE/ADMINISTRATION SECTION

Contents 11-1

ORGANIZATION CHART

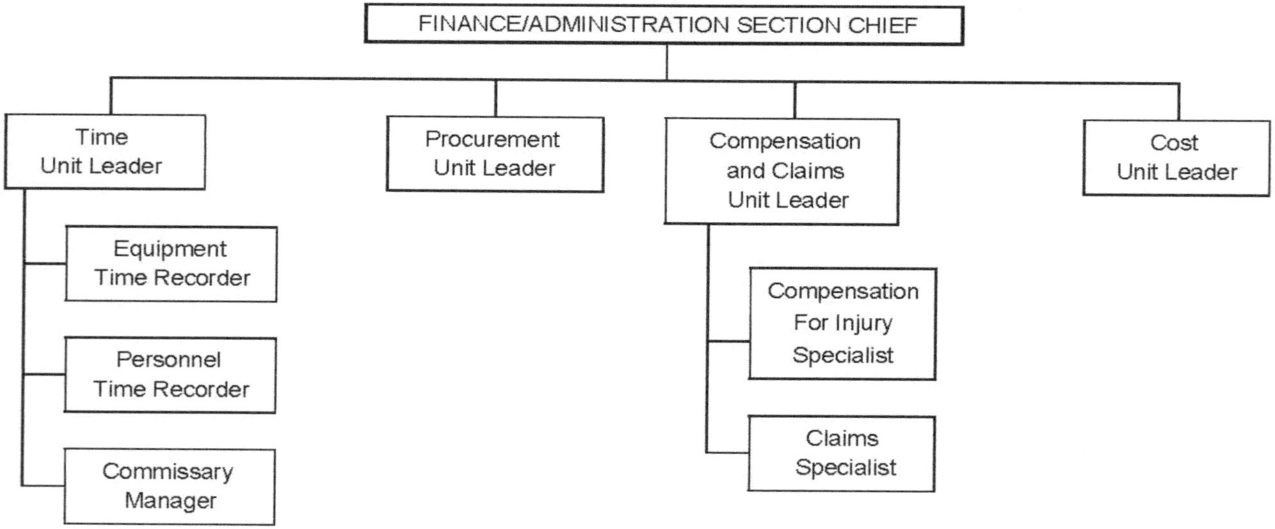

FINANCE/ADMINISTRATION SECTION CHIEF

- Time Unit Leader
  - Equipment Time Recorder
  - Personnel Time Recorder
  - Commissary Manager
- Procurement Unit Leader
- Compensation and Claims Unit Leader
  - Compensation For Injury Specialist
  - Claims Specialist
- Cost Unit Leader

POSITION CHECKLISTS

FINANCE/ADMINISTRATION SECTION CHIEF – The FSC is responsible for all financial, administrative, and cost analysis aspects of the incident and for supervising members of the Finance/Administration Section:

a. Review Common Responsibilities (Page 1-2).
b. Manage all financial aspects of an incident.
c. Provide financial and cost analysis information as requested.
d. Gather pertinent information from briefings with responsible agencies.
e. Develop an operating plan for the Finance/Administration Section; fill supply and support needs.
f. Determine need to set up and operate an incident commissary.
g. Meet with assisting and cooperating Agency Representatives as needed.
h. Maintain daily contact with agency(ies) administrative headquarters on Finance/Administration matters.
i. Ensure that all personnel time records are accurately completed and transmitted to home agencies, according to policy.
j. Provide financial input to demobilization planning.
k. Ensure that all obligation documents initiated at the incident are properly prepared and completed.
l. Brief agency administrative personnel on all incident-related financial issues needing attention or follow-up prior to leaving incident.
m. Maintain Unit/Activity Log (ICS Form 214).

TIME UNIT LEADER – The TIME is responsible for equipment and personnel time recording and for managing the commissary operations:

a. Review Common Responsibilities (Page 1-2).
b. Review Unit Leader Responsibilities (Page 1-3).
c. Determine incident requirements for time recording function.
d. Contact appropriate agency personnel/representatives.
e. Ensure that daily personnel time recording documents are prepared and in compliance with agency policy.
f. Maintain separate logs for overtime hours.
g. Establish commissary operation on larger or long-term incidents as needed.
h. Submit cost estimate data forms to Cost Unit as required.
i. Maintain records security.
j. Ensure that all records are current and complete prior to demobilization.
k. Release time reports from assisting agency personnel to the respective Agency Representatives prior to demobilization.
l. Brief Finance/Administration Section Chief on current problems and recommendations, outstanding issues, and follow-up requirements.
m. Maintain Unit/Activity Log (ICS Form 214).

EQUIPMENT TIME RECORDER – Under supervision of the Time Unit Leader, EQTR is responsible for overseeing the recording of time for all equipment assigned to an incident:

a. Review Common Responsibilities (Page 1-2).
b. Set up Equipment Time Recorder function in location designated by Time Unit Leader.

c. Advise Ground Support Unit, Facilities Unit, and Air Support Group of the requirement to establish and maintain a file for maintaining a daily record of equipment time.

d. Assist units in establishing a system for collecting equipment time reports.

e. Post all equipment time tickets within four hours after the end of each operational period.

f. Prepare a use and summary invoice for equipment (as required) within twelve (12) hours after equipment arrival at incident.

g. Submit data to Time Unit Leader for cost effectiveness analysis.

h. Maintain current posting on all charges or credits for fuel, parts, services and commissary.

i. Verify all time data and deductions with owner/operator of equipment.

j. Complete all forms according to agency specifications.

k. Close out forms prior to demobilization.

l. Distribute copies per agency and incident policy.

m. Maintain Unit/Activity Log (ICS Form 214).

PERSONNEL TIME RECORDER - Under supervision of the Time Unit Leader, PTRC is responsible for overseeing the recording of time for all personnel assigned to an incident:

a. Review Common Responsibilities (Page 1-2).

b. Establish and maintain a file for employee time reports within the first operational period.

c. Initiate, gather, or update a time report from all applicable personnel assigned to the incident for each operational period.

d. Ensure that all employee identification information is verified to be correct on the time report.

e. Post personnel travel and work hours, transfers, promotions, specific pay provisions and terminations to personnel time documents.
f. Post all commissary issues to personnel time documents.
g. Ensure that time reports are signed.
h. Close out time documents prior to personnel leaving the incident.
i. Distribute all time documents according to agency policy.
j. Maintain a log of excessive hours worked and give to Time Unit Leader daily.
k. Maintain Unit/Activity Log (ICS Form 214).

COMMISSARY MANAGER – Under the supervision of the Time Unit Leader, CMSY is responsible for commissary operations and security:

a. Review Common Responsibilities (Page 1-2).
b. Set up and provide commissary operation to meet incident needs.
c. Establish and maintain adequate security for commissary.
d. Request commissary stock through Supply Unit Leader.
e. Maintain complete record of commissary stock including invoices for material received, issuance records, transfer records and closing inventories.
f. Maintain commissary issue record by crews and submit records to Time Recorder during or at the end of each operational period.
g. Use proper agency forms for all record keeping.
h. Complete forms according to agency specification.
i. Ensure that all records are closed out and commissary stock is inventoried and returned to Supply Unit prior to demobilization.
j. Maintain Unit/Activity Log (ICS Form 214).

PROCUREMENT UNIT LEADER – The PROC is responsible for administering all financial matters pertaining to vendor contracts, leases, and fiscal agreements:

a. Review Common Responsibilities (Page 1-2).
b. Review Unit Leader Responsibilities (Page 1-3).
c. Review incident needs and any special procedures with Unit Leaders, as needed.
d. Coordinate with local jurisdiction on plans and supply sources.
e. Obtain Incident Procurement Plan.
f. Prepare and authorize contracts and land use agreements.
g. Draft Memorandum of Understanding.
h. Establish contracts and agreements with supply vendors.
i. Provide for coordination between the Ordering Manager, agency dispatch, and all other procurement organizations supporting the incident.
j. Ensure that a system is in place that meets agency property management requirements. Ensure proper accounting for all new property.
k. Interpret contracts and agreements; resolve disputes within delegated authority.
l. Coordinate with Compensation/Claims Unit for processing claims.
m. Coordinate use of impress funds as required.
n. Complete final processing of contracts and send documents for payment.
o. Coordinate cost data in contracts with Cost Unit Leader.
p. Brief Finance/Administration Section Chief on current problems and recommendations, outstanding issues, and follow-up requirements.
q. Maintain Unit/Activity Log (ICS Form 214).

COMPENSATION/CLAIMS UNIT LEADER – The COMP is responsible for the overall management and direction of all administrative matters pertaining to compensation for injury and claims-related activities (other than injury) for an incident:

a. Review Common Responsibilities (Page 1-2).
b. Review Unit Leader Responsibilities (Page 1-3).
c. Establish contact with incident Safety Officer and Liaison Officer, or Agency Representatives if no Liaison Officer is assigned.
d. Determine the need for Compensation for Injury Specialists and Claims Specialists and order personnel as needed.
e. Establish a Compensation for Injury work area within or as close as possible to the Medical Unit.
f. Review Incident Medical Plan (ICS Form 206).
g. Review procedures for handling claims with Procurement Unit.
h. Periodically review logs and forms produced by Compensation/Claims Specialists to ensure compliance with agency requirements and policies.
i. Ensure that all Compensation for Injury and Claims logs and forms are complete and routed to the appropriate agency for post-incident processing prior to demobilization.
j. Maintain Unit/Activity Log (ICS Form 214).

COMPENSATION FOR INJURY SPECIALIST – Under the supervision of the Compensation/Claims Unit Leader, the Compensation For Injury Specialist is responsible for administering financial matters resulting from serious injuries and fatalities occurring on an incident. Close coordination is required with the Medical Unit:

a. Review Common Responsibilities (Page 1-2).
b. Collocate Compensation for Injury operations with those of the Medical Unit when possible.

c. Establish procedure with Medical Unit Leader on prompt notification of injuries or fatalities.

d. Obtain copy of Incident Medical Plan (ICS Form 206).

e. Provide written authority for persons requiring medical treatment.

f. Ensure that correct agency forms are being used.

g. Provide correct billing forms for transmittal to doctor and/or hospital.

h. Monitors and reports on status of hospitalized personnel.

i. Obtain all witness statements from Safety Officer and/or Medical Unit and review for completeness.

j. Maintain log of all injuries occurring on incident.

k. Coordinate/handle all administrative paperwork on serious injuries or fatalities.

l. Coordinate with appropriate agency (ies) to assume responsibility for injured personnel in local hospitals prior to demobilization.

m. Maintain Unit/Activity Log (ICS Form 214).

CLAIMS SPECIALIST – Under the supervision of the Compensation/Claims Unit Leader, the CLMS is responsible for managing all claims-related activities (other than injury) for an incident:

a. Review Common Responsibilities (Page 1-2).

b. Develop and maintain a log of potential claims.

c. Coordinate claims prevention plan with applicable incident functions.

d. Initiate investigation on all claims other than personnel injury.

e. Ensure that site and property involved in investigation are protected.

f. Coordinate with investigation team as necessary.

g. Obtain witness statements pertaining to claims other than personnel injury.

h. Document any incomplete investigations.
i. Document follow-up action needs by local agency.
j. Keep the Compensation/Claims Unit Leader advised on nature and status of all existing and potential claims.
k. Ensure use of correct agency forms.
l. Maintain Unit/Activity Log (ICS Form 214).

COST UNIT LEADER – The COST is responsible for collecting all cost data, performing cost effectiveness analyses, and providing cost estimates and cost saving recommendations for the incident:

a. Review Common Responsibilities (Page 1-2).
b. Review Unit Leader Responsibilities (Page 1-3).
c. Coordinate with agency headquarters on cost reporting procedures.
d. Collect and record all cost data.
e. Develop incident cost summaries.
f. Prepare resources-use cost estimates for the Planning Section.
g. Make cost-saving recommendations to the Finance/ Administration Section Chief.
h. Complete all records prior to demobilization.
i. Maintain Unit/Activity Log (ICS Form 214).

FINANCE/ADMINISTRATION SECTION PLANNING CYCLE GUIDE

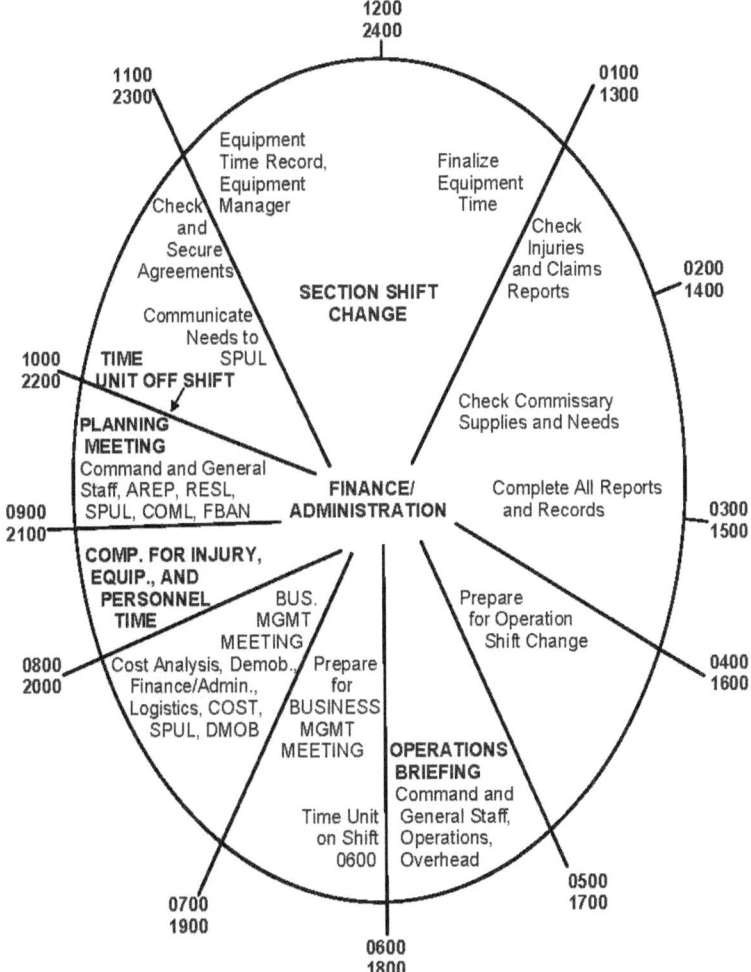

Example Based on 12-Hour Operational Period

Notes

CHAPTER 12

ORGANIZATIONAL GUIDES

Contents 12-1

FULLY ACTIVATED INCIDENT COMMAND SYSTEM ORGANIZATION CHART

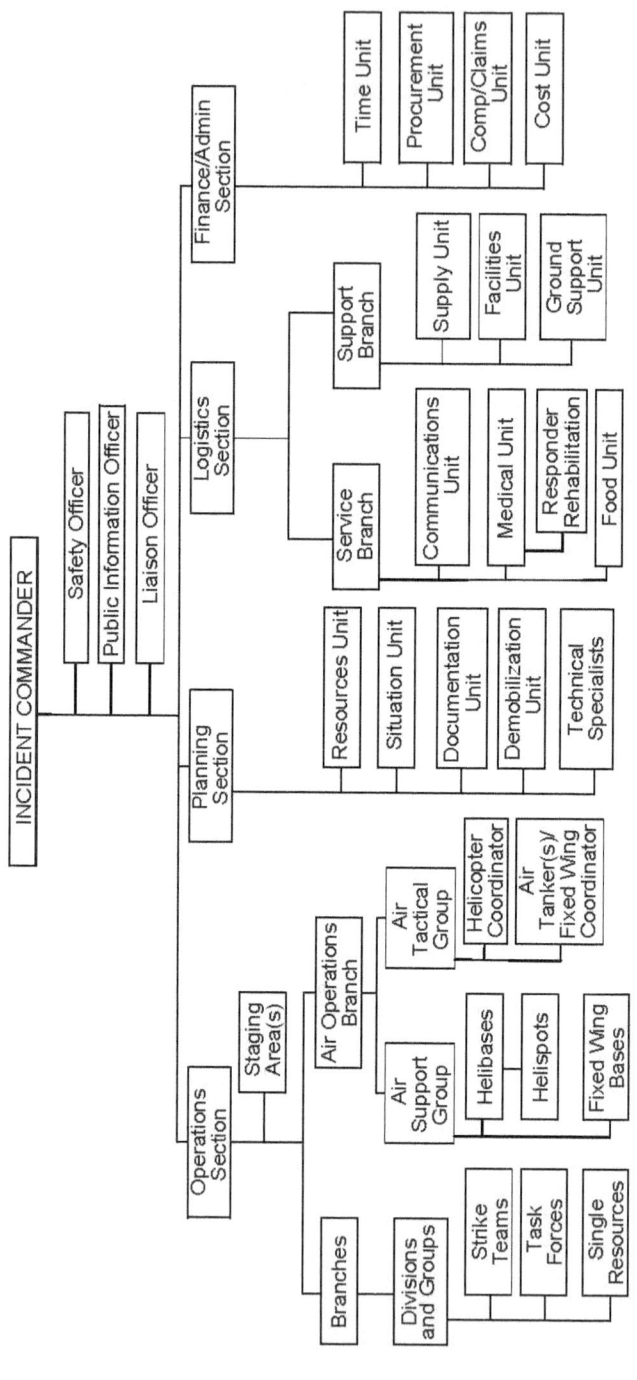

WILDLAND FIRE ORGANIZATIONAL DEVELOPMENT
INTRODUCTION

The following series of organizational charts depict examples of how the Incident Command System can be used on fires involving wildland (grass, brush, timber fuels). The charts show examples of ICS organizations for initial attack fires through incidents that grow to such size as to require very large organizational structures to manage the personnel and equipment assigned to these incidents.

Certain terms are used to identify the level of resource commitment or organizations structure. The terms associated with these levels are:

Initial Attack – This example depicts an agency's initial response level (four engines, a bulldozer, a wildland firefighting handcrew, one helicopter and one Command Officer) to a reported wildland fire and how those resources might be organized to handle the situation. At the same time, the organization is designed to rapidly expand if necessitated by fire growth.

Reinforced Response – This example depicts an expansion of the organizational structure to accommodate additional resources.

Extended Attack – This example depicts an organization that may be appropriate for incidents that may require even more resources and an extended period of time to control. The time frames for these incidents may run into multiple operational periods covering many days with enhanced logistics and planning requirements.

<u>Multi-Branch</u> – This example depicts an organization that may be used for wildland incidents that have grown in area to require multiple levels of management to accommodate span of control concerns and increased support for the number of personnel assigned to the incident.

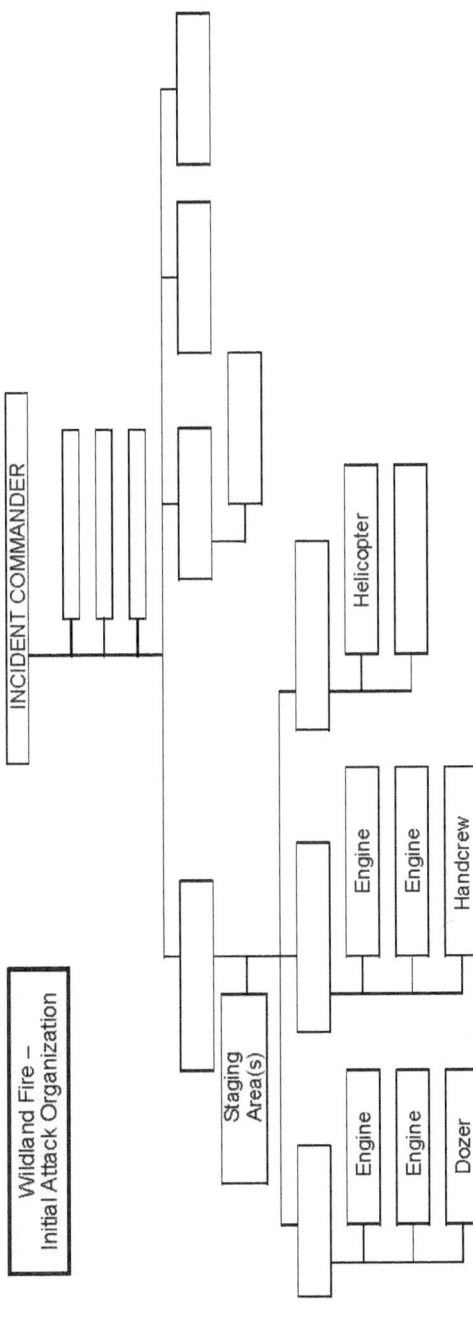

Wildland Fire –
Initial Attack Organization

INCIDENT COMMANDER

Staging Area(s)

Engine
Engine
Dozer

Engine
Engine
Handcrew

Helicopter

Wildland Fire – Initial Attack Organization (example): Initial response resources are managed by the initial Response Incident Commander (first arriving Company Officer or Command Officer) who will perform all Command and General Staff functions. Many small initial attack fires are controlled and extinguished with resource commitments at or slightly above this level. The span of control for this organization is at six to one, which is within safe guidelines of three-seven to one. Units are deployed to attack the fire with a single helicopter supporting the effort as directed by the Incident Commander. The Incident Commander has identified a Staging Area for use in the event additional resources arrive before tactical assignments for these resources are determined.

ORGANIZATIONAL

12-5

GUIDES

Wildland Fire – Reinforced Response Organization

```
                        INCIDENT COMMANDER
                        |
                    Safety Officer
        |                    |                    |
                        Resources           Logistics
                        Unit                Section
                        |
                    Helicopter

        |              |
    Staging        Division
    Area(s)        |
                   Engine
                   Dozer
                   Handcrew
                   Strike Team
                   Water
                   Tender

    Division
    |
    Engine
    Engine
    Engine
    Strike Team
    Handcrew
    Strike Team
```

Wildland Fire – Reinforced Response Organization (example): Additional resources have arrived. Span of control concerns as well as the need for tactical supervision have necessitated that the Incident Commander establish two Divisions with qualified Supervisors assigned. A Safety Officer is assigned to monitor incident operations for safety issues and to ensure corrective steps are taken. The Resources Unit is established to assist the Incident Commander with tracking resources, and a Logistics Section Chief is assigned to begin planning and implementing logistical support for the assigned resources and to plan for the support of additional resources should they be ordered.

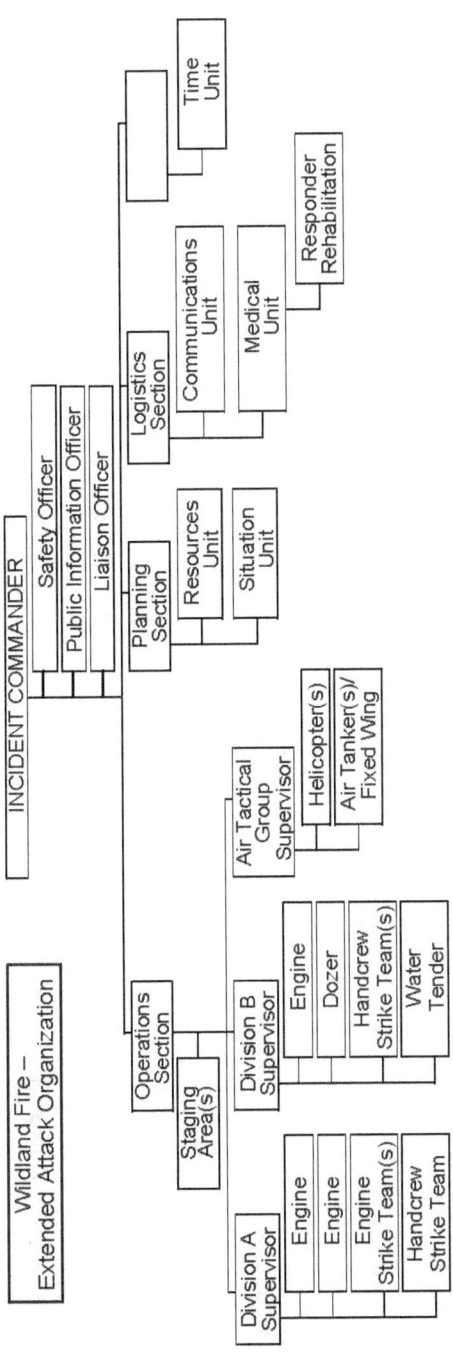

Wildland Fire – Extended Attack Organization (example): The Incident Commander has requested and received additional resources. Due to the complexity of the incident and the dynamic nature of the suppression activities, the Incident Commander has established the Operations Section Chief position. Additional aviation resources have arrived and are supervised by the Air Tactical Group Supervisor. The Incident Commander has established a Situation Unit to begin collecting incident data (mapping, weather, fire behavior predictions, etc.) to aid in the strategic and tactical planning as the incident progresses. Logistical needs have required upgraded Communications Support and a Medical Unit to handle responder injuries and rehabilitation.

ORGANIZATIONAL

GUIDES

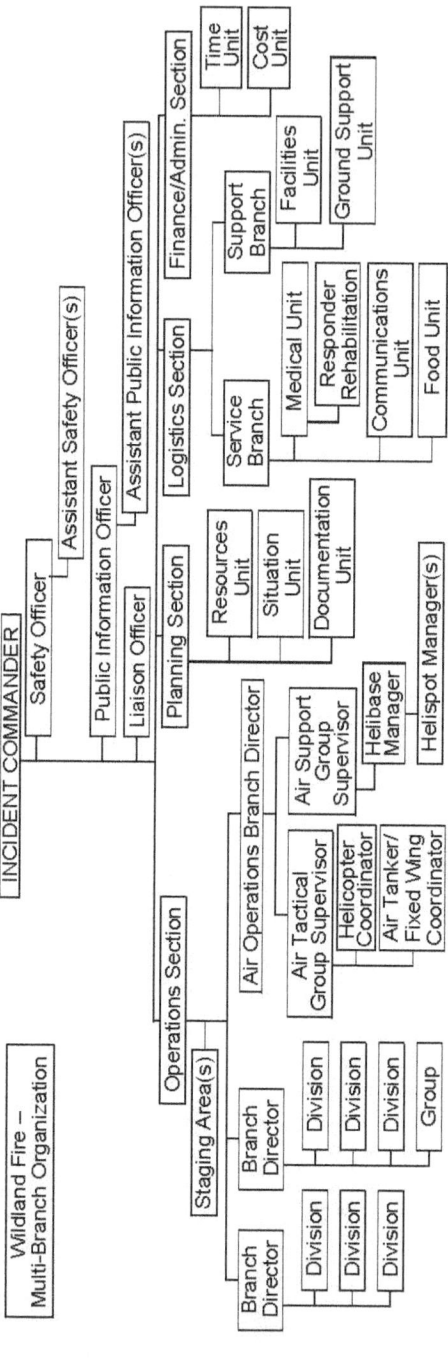

Wildland Fire – Multi-Branch Organization (example): This incident required multiple Divisions covering a large geographic area so Branches were established within the Operations Section. A full Air Operations Branch with Branch Director has been established. The Planning Section is further expanded to begin production of Incident Actions Plans for multiple Operational Periods. To ensure that adequate safety measures are taken within the expansive incident, Assistant Safety Officers have been assigned to the Safety Officer. These Assistants can be assigned to individual Branches or Divisions as well as to monitoring activities at the Base. The Command Staff is now complete to assist the Incident Commander with incident information handling and to interface with assisting and cooperating agencies.

ORGANIZATIONAL

GUIDES

STRUCTURE FIRE ORGANIZATION DEVELOPMENT
INTRODUCTION

The following series of organizational charts depict examples of how the incident Command System can be used on fires involving structures. The charts show examples of ICS organizations for initial attack fires through incidents that grow to such size as to require very large organizational structures to manage the personnel and equipment assigned to these incidents.

Certain terms are used to identify the level of resource commitment or organizations structure. The terms associated with these levels are:

Initial Attack – This example depicts an agency's initial response level (three Engines, one Truck Company, and a Command Officer) to a reported fire involving a building and how those resources might be organized to handle the situation. At the same time, the organization is designed to rapidly expand if necessitated by fire growth.

Reinforced Response – This example depicts an expansion of the organizational structure to accommodate additional resources. In this case, a second alarm has been ordered and received along with resources to assist the Incident Commander and support the personnel on scene.

Multi-Branch – This example depicts an organization that may be used for incidents that have grown in area to require multiple levels of management to accommodate span of control concerns and increased support for the number of personnel assigned to the incident.

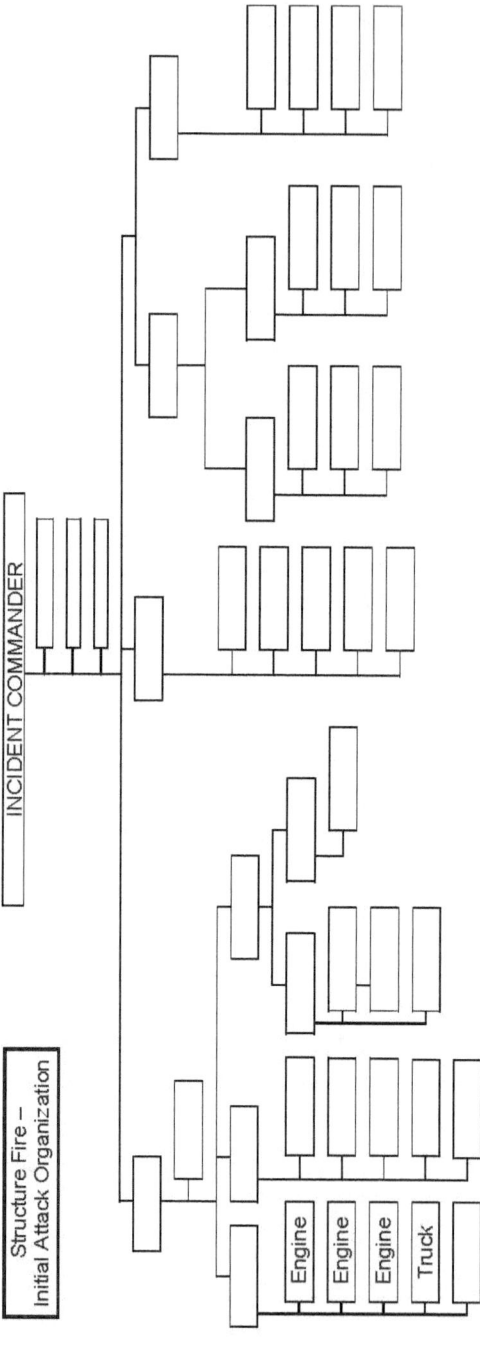

Structure Fire – Initial Attack Organization (example): This example depicts the assignment of three engines, a single truck company and a Command Officer on a structural fire. The Incident Commander manages all elements of the response. The only formal ICS position identified is that of Incident Commander. If these resources can handle the incident and no escalation is anticipated, no further ICS development is advised.

ORGANIZATIONAL

GUIDES

12-10

Structure Fire –
Reinforced Response Organization

INCIDENT COMMANDER

Safety Officer

Operations Section

Staging Area(s)

Rapid Intervention Crew/Company

Division or Group

Division or Group

Engine
Engine
Engine
Truck

Engine
Engine
Engine
Truck
Ambulance

Resources Unit

Responder Rehabilitation

Structure Fire – Reinforced Response Organization (example): Additional suppression resources have arrived and are deployed. An Operations Section Chief is activated to manage the dynamic suppression efforts. Further development of the Operations Section could include either Divisions (Division A, B, ... or Roof Division, or Division 3 for third floor operations) or Groups (Attack, Support, Rescue or Ventilation) or a combination of both (for multi-story buildings, Division 2 and 3 and a Ventilation Group may be established). The Incident Commander has activated the Safety Officer position to monitor all incident activities for safety issues and to ensure corrective actions are taken. In addition, the Incident Commander has established a Staging Area and a rapid intervention capability. The Resources Unit will assist in Resource tracking and a Responder Rehabilitation Unit is established.

ORGANIZATIONAL   GUIDES

12-11

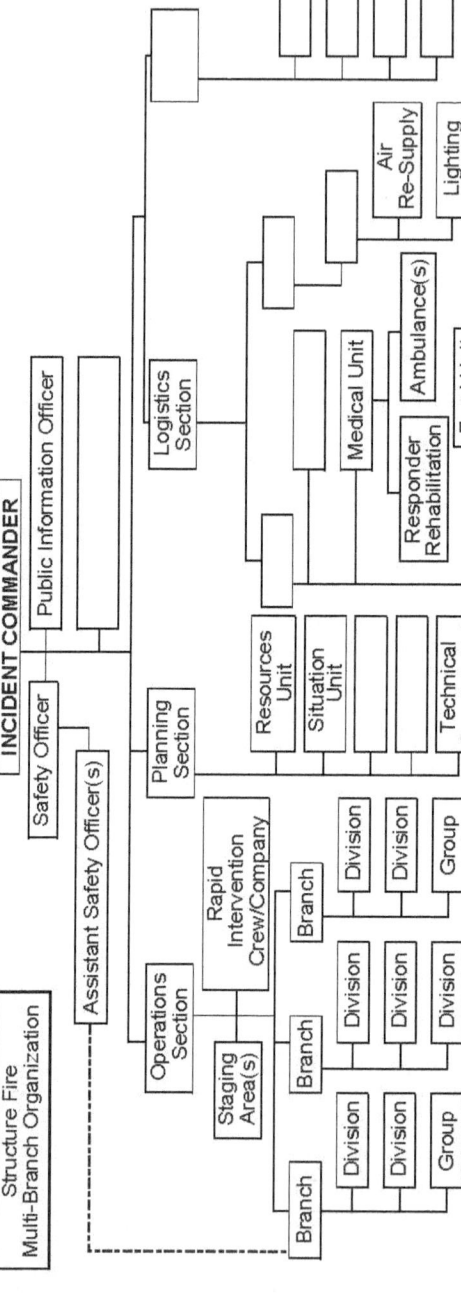

Structure Fire – Multi-Branch Organization (example): In this case, the incident is large enough that Branches have been developed and Assistant Safety Officers are assigned to either specific Branches or to individual Divisions. More elements of the Planning Section are activated as well as the Section Chief, the Situation Unit and Technical Specialists as needed. The Logistics Section is staffed with a Section Chief and elements necessary to support a long-term incident. An Information Officer is assigned to deal with inquiries from the media and local citizens.

ORGANIZATIONAL

GUIDES

ICS ORGANIZATION GUIDE

C O M M A N D

1. Incident Commander - one per incident, unless incident is multi-jurisdictional.
2. Multi-jurisdictional incidents establish Unified Command with each jurisdiction supplying an individual to represent agency in Unified Command Structure.
3. Incident Commander may have Deputy.
4. Command Staff Officer - one per function per incident.
5. Command Staff may have Assistants as needed or as required by statute or standard.
6. Agency Representatives report to Liaison Officer on Command Staff.

INCIDENT BASE RECOMMENDED MINIMUM PERSONNEL REQUIREMENTS
(PER TWELVE-HOUR OPERATIONAL PERIOD)

(If camps are established, the minimum personnel requirements for the Base may be modified or additional personnel may be added to support camps.)

UNIT POSITION	SIZE OF INCIDENT (NUMBER OF DIVISIONS)				
	2	5	10	15	25
OPERATIONS					
Operations Section Chief (OSC)	One Per Operational Period				
Branch Director (OPBD)		2	3	4	6
Division/Group Supervisor (DIVS)	2	5	10	15	25
Strike Team Leaders (STCR/DZ/EN/LM/PL)	As Needed				
Task Force Leaders (TFLD)	As Needed				
Air Operations Branch Director (AOBD)		1	1	1	1
Air Tactical Group Supervisor (ATGS)	1	1	1	1	1
Air Tanker/Fixed Wing Coordinator (ATCO)	As Needed				
Helicopter Coordinator (HLCO)	As Needed				
Air Support Group Supervisor (ASGS)	1	1	1	1	1
Helibase Manager (HEB)	One Per Helibase				
Helispot Manager (HESM)	One Per Helispot				
Fixed Wing Support Leader	One Per Airport				
Staging Area Manager (STAM)	One Per Staging Area				
Technical Specialist (THSP)	As Needed				
PLANNING					
Planning Section Chief (PSC)	One Per Incident				
Resources Unit Leader (RESL)	1	1	1	1	1
Status Recorders (SCKN)	1	2	3	3	3
Check-In Recorders (SCKN)	As Needed				
Technical Specialists (THSP)	As Needed				
Situation Unit Leader (SITL)	1	1	1	1	1
Field Observer (FOBS)		1	2	2	3
Weather Observer (WOBS)	As Needed				
GIS Specialist (GISS)	As Needed				
Damage Inspection Specialist (DINS)	As Needed				
Aerial/Ortho Photo Analyst	As Needed				
Display Processor (DPRO)		1	1	1	2
IR Equipment Operators	Two If Needed				
Computer Terminal Operator		1	1	1	1
Photographer			1	1	1
Documentation Unit Leader (DOCL)		1	1	1	1
Demobilization Unit Leader (DMOB)			1	1	1
(Demobilization Recorders from Resources)	As Needed				

	UNIT POSITION	SIZE OF INCIDENT (NUMBER OF DIVISIONS)				
		2	5	10	15	25
LOGISTICS	Logistics Section Chief (LSC)	One Per Incident				
	Service Branch Director (SVBD)	As Needed				
	Communications Unit Leader (COML)	1	1	1	1	1
	Incident Communications Manager (INCM)	1	1	1	1	1
	Message Center Operator (MCOP)		1	1	2	2
	Messenger		1	2	2	2
	Communications Technician		1	2	4	4
	Medical Unit Leader (MEDL)	1	1	1	1	1
	Medical Unit Assistant(s)	As Needed				
	Fireline EMT (FEMT)	As Needed				
	Responder Rehabilitation Manager	As Needed				
	Food Unit Leader (FDUL)		1	1	1	1
	Food Unit Assistant (each camp)	As Needed				
	Mobile Food Service		1	1	1	1
	Support Branch Director (SUBD)	As Needed				
	Supply Unit Leader (SPUL)		1	1	1	1
	Camp Supply Assistant (each camp)	As Needed				
	Ordering Manager (ODRM)			1	1	1
	Receiving/Distribution Manager (RCDM)		1	1	1	1
	Helpers		2	2	2	2
	Facility Unit Leader (FACL)		1	1	1	1
	Base Manager (BCMG)		1	1	1	1
	Camp Manager (each camp) (BCMG)	As Needed				
	Facility Maintenance Specialist (FMNT)		1	1	1	1
	Security Manager (SECM)		1	1	1	1
	Helpers		6	6	12	12
	Ground Support Unit Leader (GSUL)	1	1	1	1	1
	Equipment Manager (EQPM)		1	1	1	1
	Ground Support Assistant(s)	As Needed				
	Equipment Timekeeper		1	1	1	1
	Mechanics	1	1	3	5	7
	Drivers	As Needed				
	Operators	As Needed				
FIN-ADMIN	Finance/Administration Section Chief (FSC)	One Per Incident				
	Time Unit Leader (TIME)		1	1	1	1
	Time Recorder, Personnel (PTRC)		1	3	3	5
	Time Recorder, Equipment (EQTR)		1	2	2	3
	Procurement Unit Leader (PROC)		1	1	1	1
	Compensation/Claims Unit Leader (COMP)		1	1	1	1
	Compensation For Injury Specialist	As Needed				
	Claims Specialist (CLMS)	As Needed				
	Cost Unit Leader (COST)		1	1	1	1
	Cost Analyst			1	1	1
	Technical Specialist (THSP)	As Needed				

T-CARD COLORS AND USES

Ten different color resource cards (T-cards) are used to denote kind of resources. The card colors and resources they represent are:

KIND RESOURCE	CARD COLOR	FORM NUMBER
Engines	Rose	219-3
Handcrews	Green	219-2
Dozers	Yellow	219-7
Aircraft	Orange	219-6
Helicopter	Blue	219-4
Misc. Equip/Task Forces	Tan	219-8
Personnel	White	219-5
Location Labels	Gray	219-1
Property Record	White/red	219-9
Transfer Tag	White Tag	219-9A

INCIDENT COMMAND SYSTEM FORMS

Forms and records that are routinely used in the ICS are listed below. Those marked with an (*) are commonly used in written Incident Action Plans.

	Incident Briefing	ICS Form 201
*	Objectives	ICS Form 202
*	Organization Assignment List	ICS Form 203
*	Assignment List	ICS Form 204
*	Incident Radio Communications Plan	ICS Form 205
*	Medical Plan	ICS Form 206
	Incident Organization Chart	ICS Form 207
	Site Safety and Control Plan	ICS Form 208
	Incident Status Summary	ICS Form 209
	Check-In List	ICS Form 211
	Demobilization Vehicle Safety Inspection	ICS Form 212
	General Message	ICS Form 213
	Unit/Activity Log	ICS Form 214
	Incident Safety Analysis – Generic/Wildland	ICS Form 215 AG/AW
	Operational Planning Worksheet – Generic/Wildland	ICS Form 215 G/W
	Incident Resource Projection Matrix	ICS Form 215 M
	Radio Requirements Worksheet	ICS Form 216
	Support Vehicle Inventory	ICS Form 218
	Resource Status Card (1-9A)	ICS Form 219
	Air Operations Summary	ICS Form 220
	Demobilization Checkout	ICS Form 221
	Incident Weather Forecast Request	ICS Form 222
	Tentative Release List	ICS Form 223
	Crew Performance Rating	ICS Form 224
	Incident Personnel Performance Rating	ICS Form 225
	Compensation for Injury Log	ICS Form 226
	Claims Log	ICS Form 227
	Contractor/Vendor Performance Evaluation	ICS Form 230

ICS MAP DISPLAY SYMBOLOGY

SUGGESTED FOR PLACEMENT ON BASE CAMP
MINIMUM RECOMMENDED

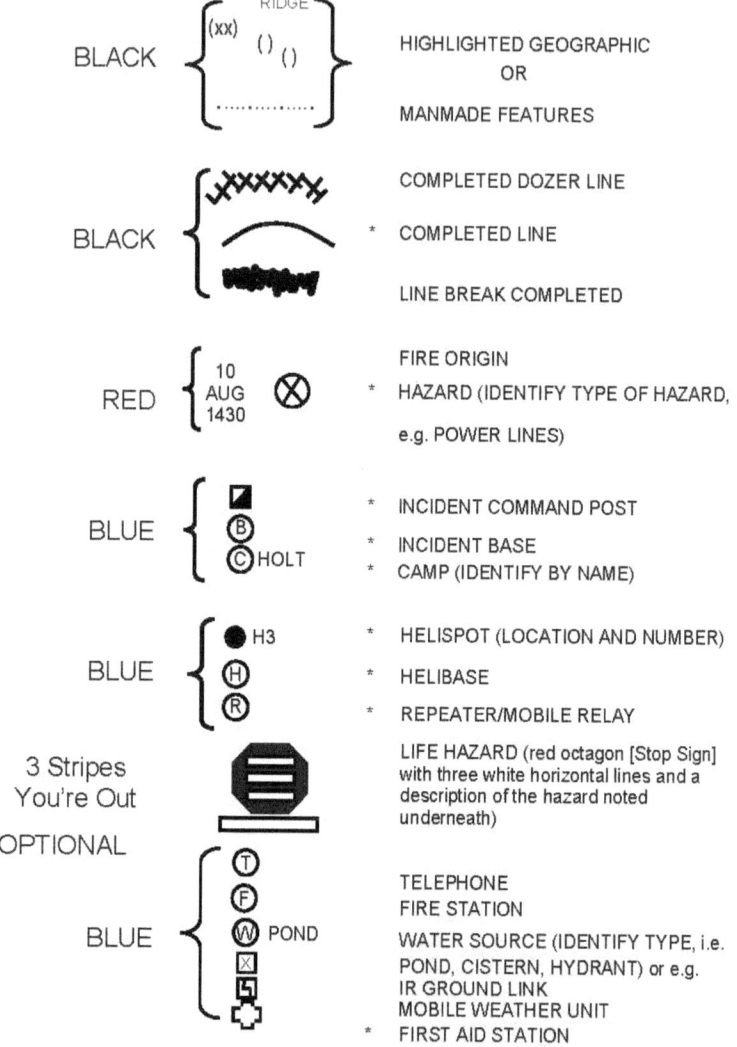

BLACK — HIGHLIGHTED GEOGRAPHIC
OR
MANMADE FEATURES

BLACK — COMPLETED DOZER LINE
* COMPLETED LINE
LINE BREAK COMPLETED

RED — FIRE ORIGIN
* HAZARD (IDENTIFY TYPE OF HAZARD, e.g. POWER LINES)

BLUE — * INCIDENT COMMAND POST
* INCIDENT BASE
* CAMP (IDENTIFY BY NAME)

BLUE — * HELISPOT (LOCATION AND NUMBER)
* HELIBASE
* REPEATER/MOBILE RELAY

3 Stripes
You're Out

OPTIONAL

LIFE HAZARD (red octagon [Stop Sign] with three white horizontal lines and a description of the hazard noted underneath)

BLUE — TELEPHONE
FIRE STATION
WATER SOURCE (IDENTIFY TYPE, i.e. POND, CISTERN, HYDRANT) or e.g.
IR GROUND LINK
MOBILE WEATHER UNIT
* FIRST AID STATION

* - TO BE USED ON INCIDENT BRIEFING AND ACTION PLAN MAPS (NO COLOR)

ICS MAP DISPLAY SYMBOLOGY (Continued)

SUGGESTED FOR PLACEMENT ON OVERLAYS

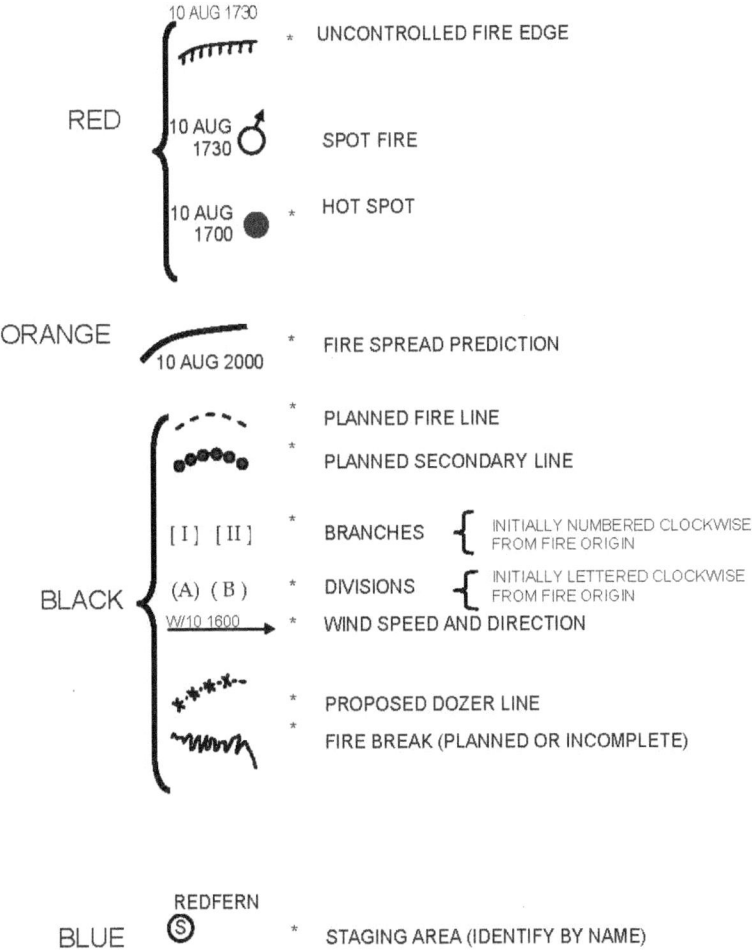

RED

10 AUG 1730 — * UNCONTROLLED FIRE EDGE

10 AUG 1730 — * SPOT FIRE

10 AUG 1700 — * HOT SPOT

ORANGE — * FIRE SPREAD PREDICTION
10 AUG 2000

BLACK

* PLANNED FIRE LINE

* PLANNED SECONDARY LINE

[I] [II] — * BRANCHES — INITIALLY NUMBERED CLOCKWISE FROM FIRE ORIGIN

(A) (B) — * DIVISIONS — INITIALLY LETTERED CLOCKWISE FROM FIRE ORIGIN

W/10 1600 — * WIND SPEED AND DIRECTION

* PROPOSED DOZER LINE

* FIRE BREAK (PLANNED OR INCOMPLETE)

BLUE

REDFERN
S — * STAGING AREA (IDENTIFY BY NAME)

ALL OVERLAYS MUST CONTAIN REGISTRATION MARKS.
THESE MAY CONSIST OF IDENTIFIED ROAD INTERSECTIONS,
TOWNSHIP/RANGE COORDINATES, MAP CORNERS, ETC.

RESOURCES UNIT FUNCTIONS AND INTERACTIONS

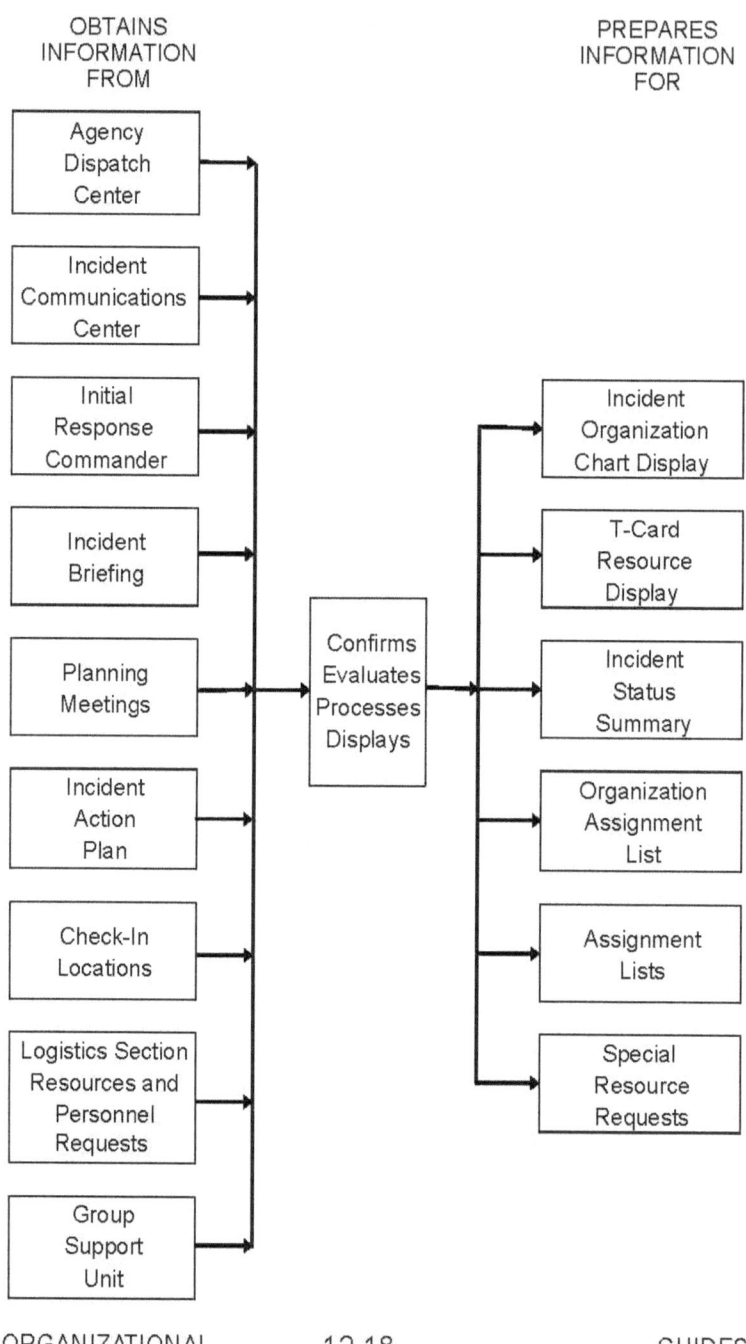

OBTAINS INFORMATION FROM

Agency Dispatch Center

Incident Communications Center

Initial Response Commander

Incident Briefing

Planning Meetings

Incident Action Plan

Check-In Locations

Logistics Section Resources and Personnel Requests

Group Support Unit

Confirms Evaluates Processes Displays

PREPARES INFORMATION FOR

Incident Organization Chart Display

T-Card Resource Display

Incident Status Summary

Organization Assignment List

Assignment Lists

Special Resource Requests

SITUATION UNIT FUNCTIONS AND INTERACTIONS

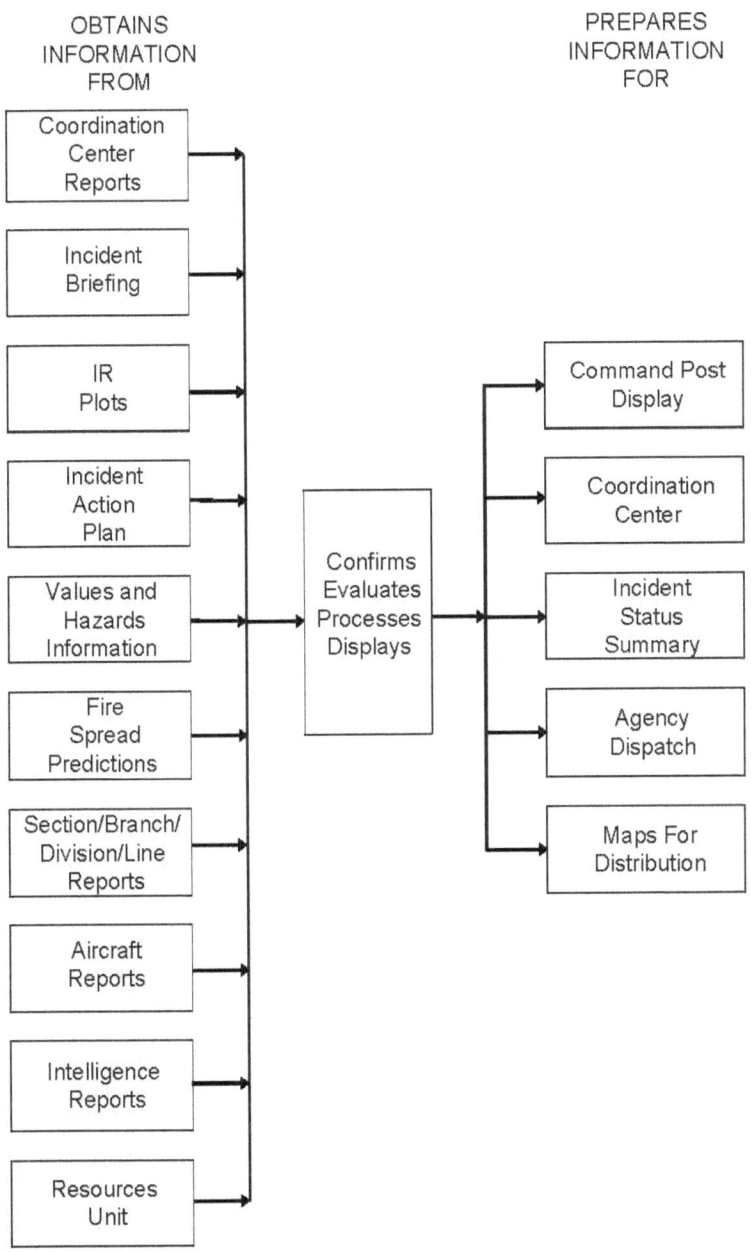

OBTAINS INFORMATION FROM

PREPARES INFORMATION FOR

Coordination Center Reports

Incident Briefing

IR Plots

Incident Action Plan

Values and Hazards Information

Fire Spread Predictions

Section/Branch/ Division/Line Reports

Aircraft Reports

Intelligence Reports

Resources Unit

Confirms Evaluates Processes Displays

Command Post Display

Coordination Center

Incident Status Summary

Agency Dispatch

Maps For Distribution

RESOURCE STATUS CHANGE REPORTING

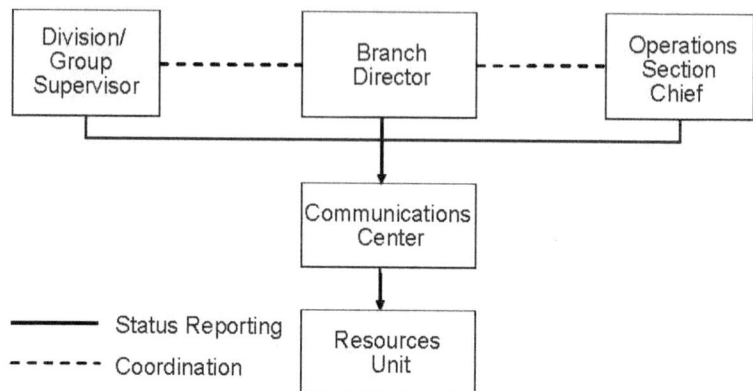

1. Report:
   A) Resources changing status (assigned, available, out of service)
   B) Resources moving between Divisions

2. Note: Authority who approves the status change is responsible for
   reporting it to Resources Unit

STRIKE TEAM LEADER INTERACTIONS

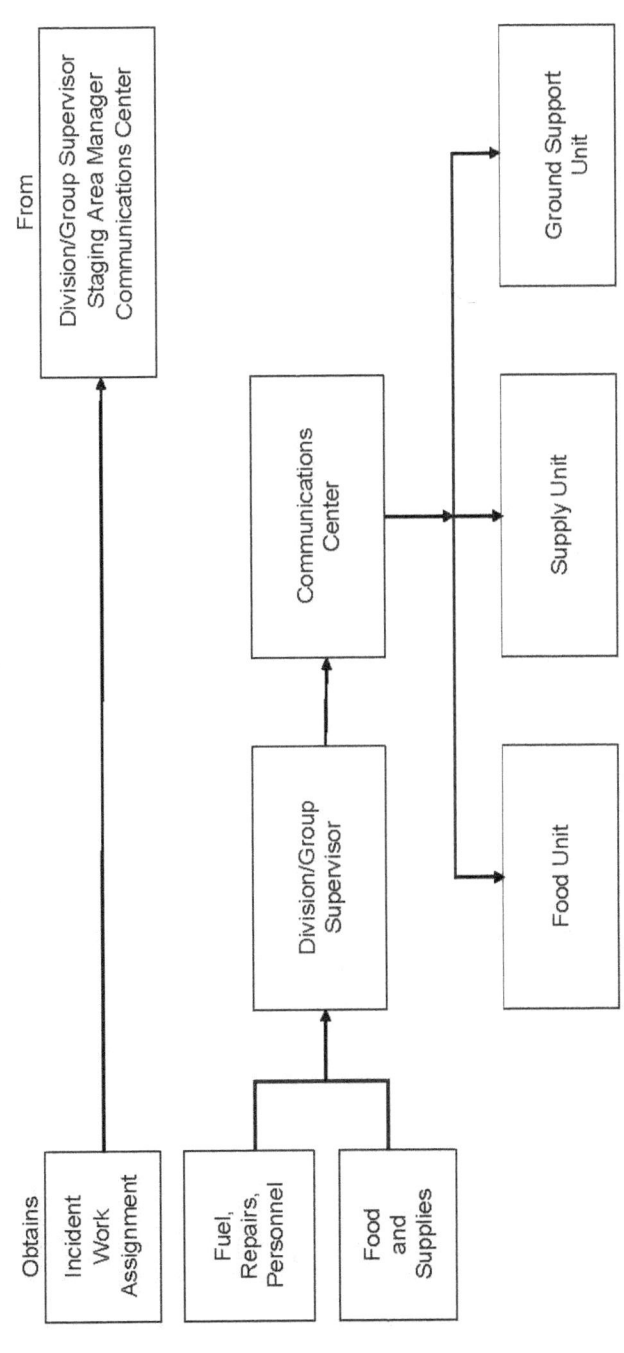

Note: Out-of-service resources interact directly with appropriate units for service and support

REASSIGN/RELEASE OF RESOURCES

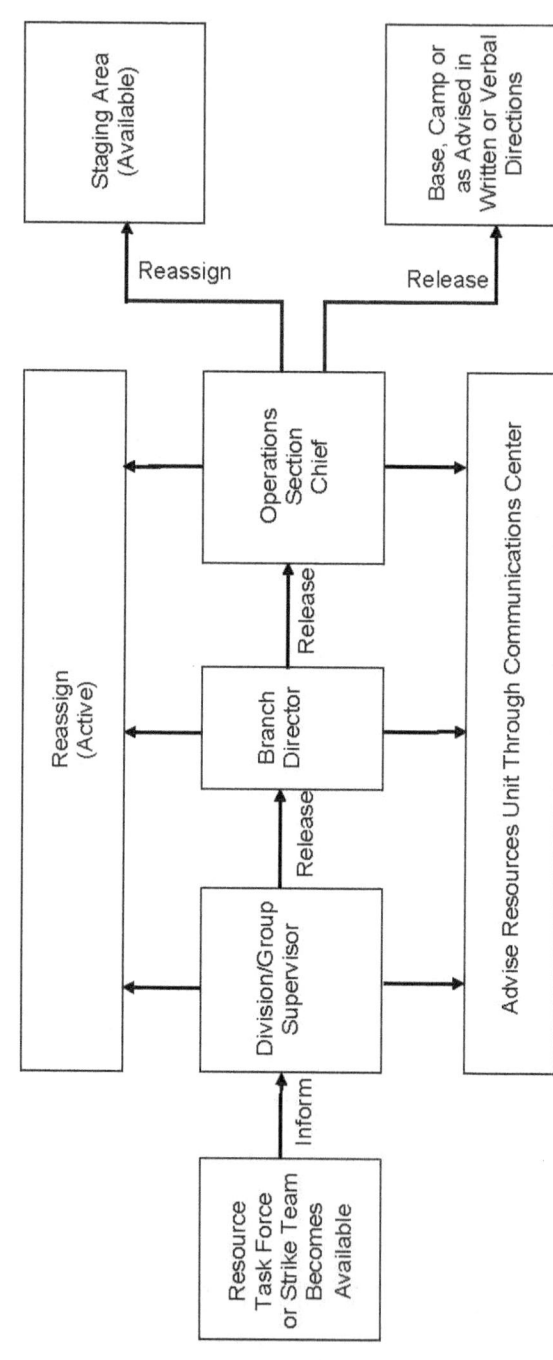

NOTE: Authority who approves the status change is responsible for reporting it to Resources Unit.

ORGANIZATIONAL   GUIDES

CAMP ORGANIZATION AND REPORTING RELATIONSHIPS

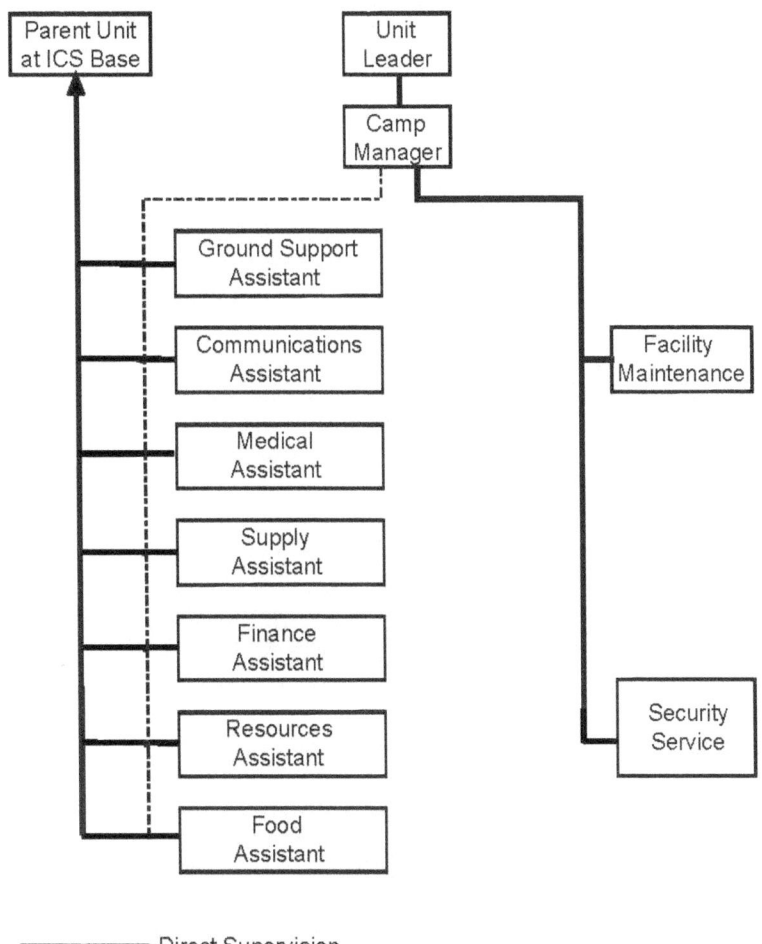

━━━━━━━━ Direct Supervision

----------- In-Camp Coordinator

The Camp Manager will provide direct supervision for all facility maintenance and security services at the Camp. Several of the functional unit activities that are performed at the Base may also be performed at the Camp(s). These functional units assigned to the Camp(s) will receive their direct supervision from their Unit Leaders at the Base. During the time that a Camp is established, the Camp Manager will be responsible to provide non-technical coordination for all Units operating within the Camp in order to ensure orderly and harmonious operation of the Camp and efficient use of all resources and personnel assigned to the Camp.

Notes

CHAPTER 13

RESOURCE TYPES AND MINIMUM STANDARDS

PRIMARY MOBILE SUPPRESSION RESOURCES
(Minimum ICS Standards)

RESOURCE	RADIO CALL	COMPONENTS	TYPES			
			1	2	3	4
Engine Company	Engine Telesquirt*	Pump Water Tank Hose 2 1/2" Hose 1 1/2" Hose 1" Ladder Master Stream Personnel	1,000 GPM 400 Gal. 1,200 Ft. 400 Ft. 200 Ft. 20 Ft. Ext. 500 GPM 4	500 GPM 400 Gal. 1,000 Ft. 500 Ft. 300 Ft. 20 Ft. Ext. - 3	120 GPM 300 Gal. - 1,000 Ft. 800 Ft. - - 3	50 GPM 200 Gal. - 300 Ft. 800 Ft. - - 3
*Engine with elevated stream capability, specify when requested.						
Truck Company	Truck	Aerial (Specify platform or ladder), Elevated Stream Ground Ladders, Personnel	75 Ft. 500 GPM 115 Ft. 4	50 Ft. 500 GPM 115 Ft. 4		
Water Tender	Water Tender	Pump Water Tank	300 GPM 2,000 Gal.	120 GPM 1,000 Gal.	50 GPM 1,000 Gal.	
Brush Patrol	Patrol	Pump - 15 GPM Hose 1" - 150 Ft. Tank - 75 Gal. Personnel - 1				
Medical/Non Transport	Rescue, Squad, Medic Engine	Non Transport, capability and personnel determined by local EMS authority	ALS	BLS		
Medical/ Transport	Ambulance, Medic	Transport, capability and personnel determined by local EMS authority	ALS	BLS		
Bulldozer	Dozer	Size Horse Power Operator Example(s)	Heavy 200 HP 1 D-7, D-8	Medium 100 HP 1 D-5, D-6	Light 50 HP 1 D-4	
Bulldozer Tender	Dozer Tender	Fuel - 100 Gal.				

PRIMARY MOBILE SUPPRESSION RESOURCES (continued)

RESOURCE	RADIO CALL	COMPONENTS	TYPE 1	TYPE 2
Hand Crew	Crew #	*Personnel, Equipment, and Transportation	• Highest training level • No use restriction • Fully mobilized • Highest experience level • Fully equipped • Permanently assigned supervision	• Minimum training or • Some use restrictions or • Not fully mobilized or • Moderate experience or • Minimum equipment or • No assigned supervision
*Indicates minimum number of crew personnel including supervision			State CDC (12) CYA (12) CCC (12) Fly Crew (8) Local Govt. Inmate (12) Paid (10)	Federal Hotshot (18) Regular (18) Fly Crew (10) Fly Crew (8) Hotshot (18)	Federal (Blue Card) (18) State (12)

RESOURCE	RADIO CALL	COMPONENTS	TYPES			
			1	2	3	4
Fire Boat	Boat	Pumping Capability	5,000 GPM	1,000 GPM	250 GPM	
Foam Tender	Foam	Class B Foam Specify: % Concentrate (1%, 3%, etc.)	500 Gal.	250 Gal.		
Air Tanker	Tanker	Gallons Examples:	3,000 C-130, P-3	1,800 SP2H, P2V	800 S-2T	200 SEAT
Helicopters	Copter	Seats, including pilot Card weight capacity (lbs) Gallons Examples	16 5,000 700 Bell 214	10 2,500 300 Bell 204, 205, 212	5 1,200 100 Bell 206	3 600 75 Hiller 12E 3T
Helitanker	Helitanker	- Fixed Tank - Air tanker Board Certified - 1,000 Minimum Gallon Capacity				
Helicopter Tender	Helitender	Fuel Equipment				
Helitack Crew	Helitack	Personnel (3) Equipment Transportation				
Aircraft Rescue Firefighting (ARFF)	ARFF	Class B Foam w/ proportioner and pump				

SUPPORT RESOURCES

RESOURCE	RADIO CALL	COMPONENTS	TYPES		
			1	2	3
Breathing Apparatus Support	Breathing Support	Filling Capability	Compressor	Cascade	
Crew Transport	Crew Transport	Passengers	30	20	10
Field Mobile Mechanic	Repair	Repair Capability	Heavy Equipment	Light Equipment	
Food Dispenser Unit	Food Dispenser	Servings/Meal	150	50	
Mobile Kitchen Unit	Mobile Kitchen	Servings/Meal	1,000	300	
Fuel Tender	Fuel Tender	Fuel Specify: Gas, Jet Fuel, Diesel, Etc.	1,000 Gal.	100 Gal.	
Heavy Equipment Transport	Transport	Capacity Examples:	Heavy D-7, D-8	Medium D-6	Light D-4
Illumination Unit	Light	Lighting Units (500 Watts each) Extension Cord Specify: Mounted or Portable	6 1,000 Ft.	3 500 Ft.	
Mobile Communication	Comm	•Consoles/ Workstations •Frequency Capability •Power Source •Telephone Systems •Personnel	2 Multi Range*, Programm-able Internal 6Trunk/16 Extension Lines 2	2 Multi Range*, Programm-able Internal 2	1 Single Range**, Programm-able External 1
* Multi Range: 150-174 MHz, 450-470 MHz, 800 MHz (Simplex and Repeated) **Single Range: 150-174 MHz only					
Portable Pump	N/A	Pumping Capacity	500 GPM	250 GPM	50 GPM
Portable Repeater	N/A	Frequency Capability*			
*When requesting resource, need to specify frequency requirements.					
Power Generator	N/A	Wattage Capacity Specify: Mounted or Portable	10,000 watts	3,000 watts	
Refrigeration Unit	Refer	Box Length (Ft.)	24	12	
Utility Transport	Utility		Over 1 Ton	1 Ton and Under	

STRIKE TEAM TYPES AND MINIMUM STANDARDS

Kind	Strike Team Types	Number/Type	Minimum Equipment Standards							Minimum Personnel		
			Pump Capacity	Water Capacity	2 1/2" Hose	1 1/2" Hose	1" Hose	Ladder	Master Stream	Strike Team Leader	Per Single Resource	Total Personnel
ENGINES	A	5 – Type 1	1,000 GPM	400 Gallons	1,200 Feet	400 Feet	200 Feet	20 Ft. Ext.	500 GPM	1	4	21
	B	5 – Type 2	500 GPM	400 Gallons	1,000 Feet	500 Feet	300 Feet	20 Ft. Ext.	N/A	1	3	16
	C	5 – Type 3	120 GPM	300 Gallons	N/A	1,000 Feet	800 Feet	N/A	N/A	1	3	16
	D	5 – Type 4	50 GPM	200 Gallons	N/A	300 Feet	800 Feet	N/A	N/A	1	3	16
CREWS	G	Handcrew combinations consisting of a minimum of 29 persons (Do not mix type 1 and Type 2 crews)	Type 1 Handcrews have no restrictions on use							1	N/A	30
	H		Type 2 Handcrews may have use restrictions							1	N/A	30
DOZERS	K	2 – Type 1 / 1 – Dozer Tender	Heavy Dozer Minimum 200 HP (D-7, D-8 or equivalent)							1	1 / 1	4
	L	2 – Type 2 / 1 – Dozer Tender	Medium Dozer Minimum 100 HP (D-5, D-6 or equivalent)							1	1 / 1	4
	M	2 – Type 3 / 1 – Dozer Tender	Light Dozer Minimum 50 HP (D-4 or equivalent)							1	1 / 1	4

RESOURCE TYPES

MINIMUM STANDARDS

Notes

CHAPTER 14

HAZARDOUS MATERIALS

Contents 14-1

INTRODUCTION

The Hazardous Materials organizational module is designed to provide an organizational structure that will provide necessary supervision and control for the essential functions required at virtually all Hazardous Materials incidents. This is based on the premise that controlling the tactical operations of companies and movement of personnel and equipment will provide a greater degree of safety and also reduce the probability of spreading of contaminants. The Hazardous Materials Group Supervisor or the Hazardous Materials Branch Director (if activated) will direct primary functions, and all resources that have a direct involvement with the hazardous material will be supervised by one of the functional leaders or the Hazardous Materials Group Supervisor.

MODULAR DEVELOPMENT

A series of examples of modular development are included to illustrate one method of expanding the incident organization:

Initial Response - The Incident Commander manages all initial response resources as well as all Command and General Staff responsibilities.

Reinforced Response - In addition to the initial response, the responsible agencies have met and established Unified Command. The Unified Incident Commanders have met and have established Unified Command. They have established a Hazardous Materials Group to manage all activities around the Control Zones and have organized Law Enforcement units into a task force to isolate the operational area. The Incident Commanders have decided to establish a Planning Section, a Staging Area, and a Safety Officer.

<u>Multi-Division/Group</u> – The Incident Commanders have activated most Command and General Staff positions and have established a combination of divisions and groups.

<u>Multi-Branch</u> – The Incident Commanders have activated all Command and General Staff positions, and have established four branches in the Operations Section.

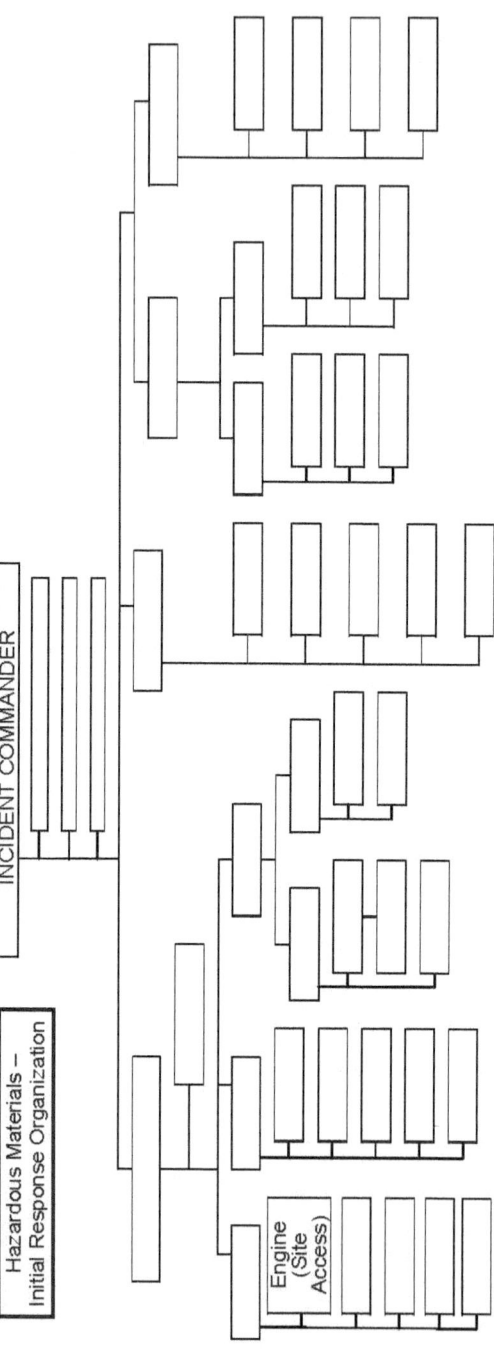

Hazardous Materials – Initial Response Organization (example): The Engine Company has arrived to find a release of a Hazardous Materials and is initiating immediate actions to isolate the area (Site Access). In addition, the Company Officer has assumed Incident Command and is ordering additional resources.

14-4

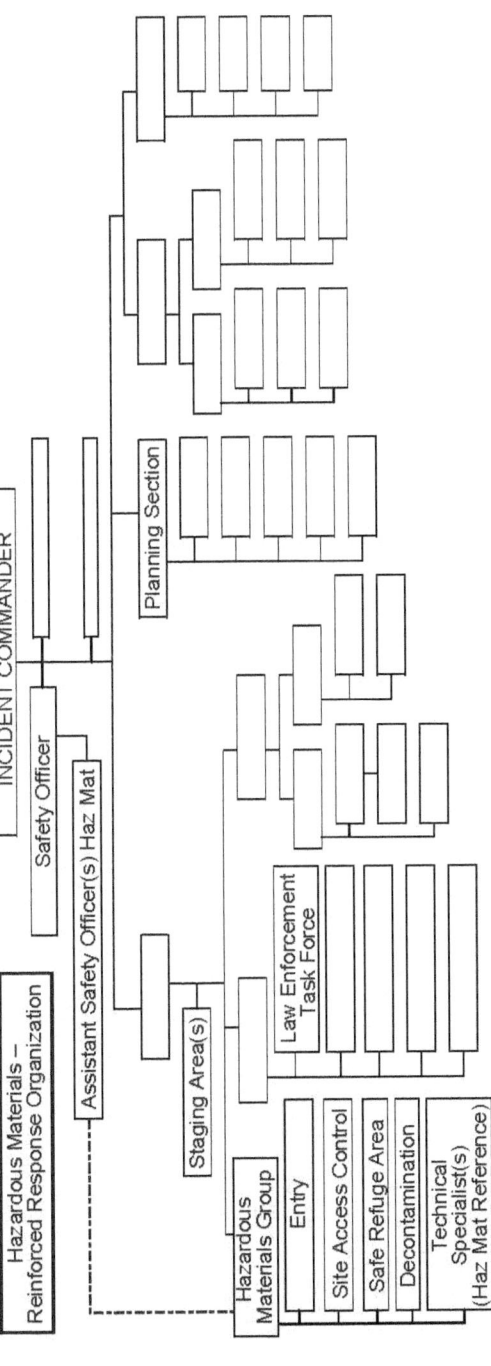

Hazardous Materials – Reinforced Response Organization (example): The Hazardous Materials response has been reinforced and a Hazardous Materials Group has been established to deal with the release. Law Enforcement responsibilities of scene security and crowd control will be assessed and handled by a Law Enforcement Group. The Planning Section Chief will accomplish initial planning and resource tracking.

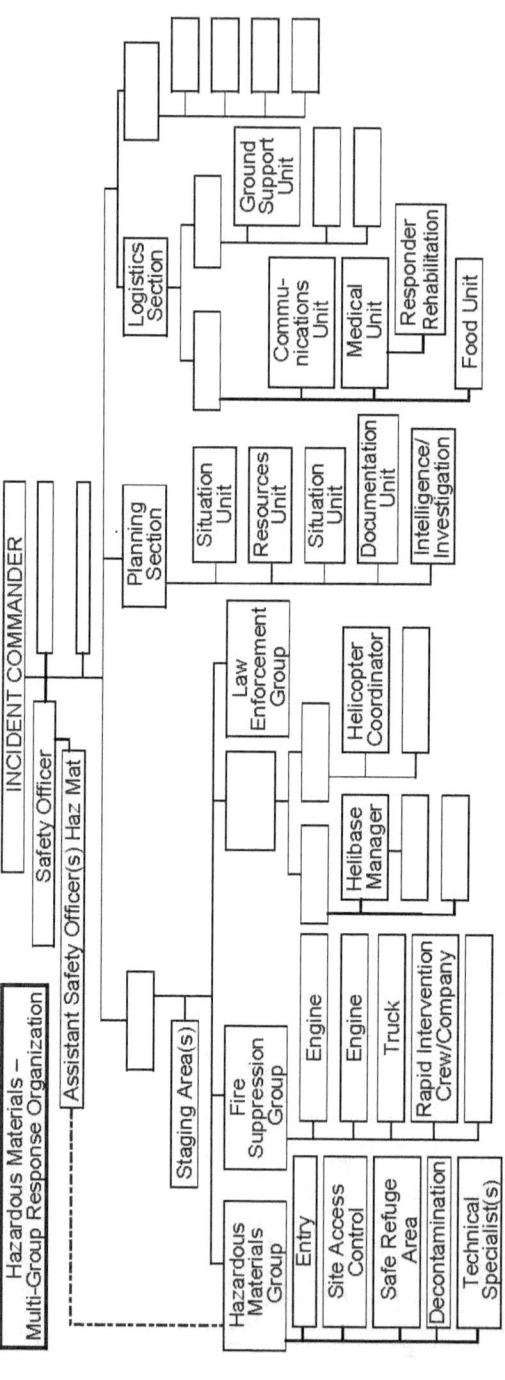

Hazardous Materials – Multi-Group Response Organization (example): Additional resources have arrived and the Incident Commander has established a Fire Suppression Group to address other risks on the incident. Aviation resources are assigned and appropriate supervision is established. Planning and Logistics Sections are partially established. An Assistant Safety Officer is specifically assigned to the Hazardous Materials Group.

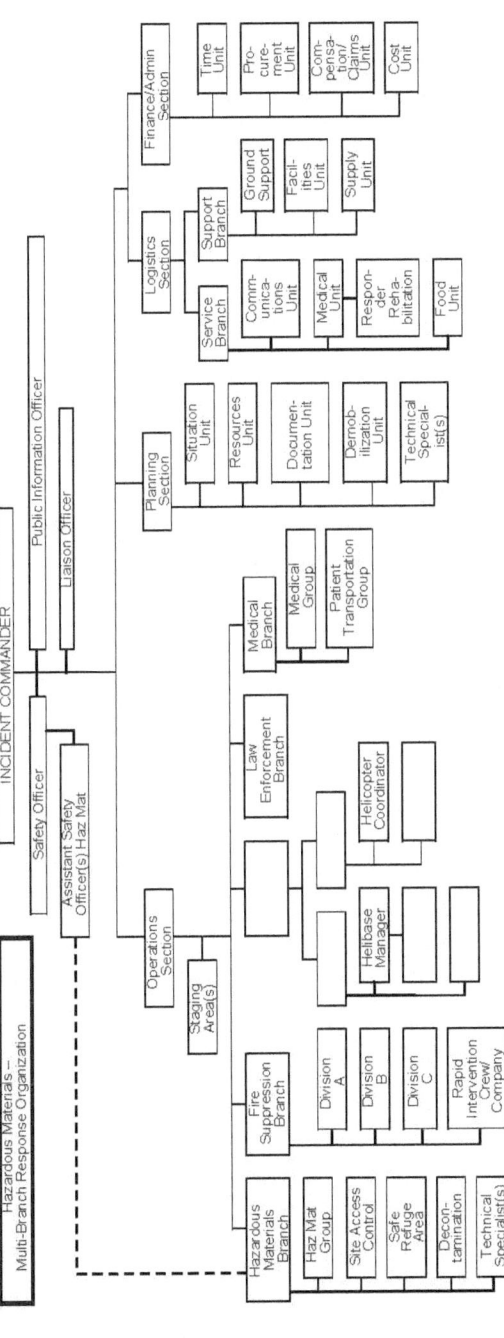

Hazardous Materials – Multi-Branch Response (example): In this case, the incident now includes more than just a Hazardous Materials release. Therefore, the complexity of the incident requires an Operations Section Chief be assigned as well as the balance of the Command and General Staff positions. Operational control is now enhanced by the assignment of Branch Directors.

POSITION CHECKLISTS

HAZARDOUS MATERIALS GROUP SUPERVISOR - The Hazardous Materials Group Supervisor or Hazardous Materials Branch Director reports to the Operations Section Chief. The Hazardous Materials Group Supervisor is responsible for the implementation of the phases of the Incident Action Plan dealing with the Hazardous Materials Group operations. The Hazardous Materials Group Supervisor is responsible for the assignment of resources within the Hazardous Materials Group, reporting on the progress of control operations and the status of resources within the group. The Hazardous Materials Group Supervisor directs the overall operations of the Hazardous Materials Group:

a. Review Common Responsibilities (Page 1-2).
b. Ensure the development of Control Zones and Access Control Points and the placement of appropriate control lines.
c. Evaluate and recommend public protection action options to the Operations Chief or Branch Director (if activated).
d. Ensure that current weather data and future weather predictions are obtained.
e. Establish environmental monitoring of the hazard site for contaminants.
f. Ensure that a Site Safety and Control Plan (ICS Form 208) is developed and implemented.
g. Conduct safety meetings with the Hazardous Materials Group.
h. Participate, when requested, in the development of the Incident Action Plan.
i. Ensure that recommended safe operational procedures are followed.
j. Ensure that the proper Personal Protective Equipment is selected and used.

k. Ensure that the appropriate agencies are notified through the Incident Commander.
l. Maintain Unit/Activity Log (ICS Form 214).

ENTRY LEADER - Reports to the Hazardous Materials Group Supervisor. The Entry Leader is responsible for the overall entry operations of assigned personnel within the Exclusion Zone:

a. Review Common Responsibilities (Page 1-2).
b. Supervise entry operations.
c. Recommend actions to mitigate the situation within the Exclusion Zone.
d. Carry out actions, as directed by the Hazardous Materials Group Supervisor, to mitigate the hazardous materials release or threatened release.
e. Maintain communications and coordinate operations with the Decontamination Leader.
f. Maintain communications and coordinate operations with the Site Access Control Leader and the Safe Refuge Area Manager (if activated).
g. Maintain communications and coordinate operations with Technical Specialist-Hazardous Materials Reference.
h. Maintain control of the movement of people and equipment within the Exclusion Zone, including contaminated victims.
i. Direct rescue operations, as needed, in the Exclusion Zone.
j. Maintain Unit/Activity Log (ICS Form 214).

DECONTAMINATION LEADER - Reports to the Hazardous Materials Group Supervisor. The Decontamination Leader is responsible for the operations of the decontamination element, providing decontamination as required by the Incident Action Plan:

a.   Review Common Responsibilities (Page 1-2).
b.   Establish the Contamination Reduction Corridor(s).
c.   Identify contaminated people and equipment.
d.   Supervise the operations of the decontamination element in the process of decontaminating people and equipment.
e.   Control the movement of people and equipment within the Contamination Reduction Zone.
f.   Maintain communications and coordinate operations with the Entry Leader.
g.   Maintain communications and coordinate operations with the Site Access Control Leader and the Safe Refuge Area Manager (if activated).
h.   Coordinate the transfer of contaminated patients requiring medical attention (after decontamination) to the Medical Group.
i.   Coordinate handling, storage, and transfer of contaminants within the Contamination Reduction Zone.
j.   Maintain Unit/Activity Log (ICS Form 214).

SITE ACCESS CONTROL LEADER - Reports to the Hazardous Materials Group Supervisor. The Site Access Control Leader is responsible for the control of the movement of all people and equipment through appropriate access routes at the hazard site and ensures that contaminants are controlled and records are maintained:

a.   Review Common Responsibilities (Page 1-2).
b.   Organize and supervise assigned personnel to control access to the hazard site.
c.   Oversee the placement of the Exclusion Control Line and the Contamination Control Line.
d.   Ensure that appropriate action is taken to prevent the spread of contamination.

e. Establish the Safe Refuge Area within the Contamination Reduction Zone. Appoint a Safe Refuge Area Manager (as needed).

f. Ensure that injured or exposed individuals are decontaminated prior to departure from the hazard site.

g. Track the movement of persons passing through the Contamination Control Line to ensure that long-term observations are provided.

h. Coordinate with the Medical Group for proper separation and tracking of potentially contaminated individuals needing medical attention.

i. Maintain observations of any changes in climatic conditions or other circumstances external to the hazard site.

j. Maintain communications and coordinate operations with the Entry Leader.

k. Maintain communications and coordinate operations with the Decontamination Leader.

l. Maintain Unit/Activity Log (ICS Form 214).

ASSISTANT SAFETY OFFICER - HAZARDOUS MATERIALS

Reports to the incident Safety Officer as an Assistant Safety Officer and coordinates with the Hazardous Materials Group Supervisor or Hazardous Materials Branch Director, if activated. The Assistant Safety Officer-Hazardous Materials coordinates safety related activities directly relating to the Hazardous Materials Group operations as mandated by 29 CFR Part 1910.120 and applicable state and local laws. This position advises the Hazardous Materials Group Supervisor (or Hazardous Materials Branch Director) on all aspects of health and safety and has the authority to stop or prevent unsafe acts. It is mandatory that an Assistant Safety Officer-Hazardous Materials be appointed at all hazardous materials incidents. In a multi-activity incident the Assistant Safety Officer-Hazardous Materials does not act as the Safety Officer for the overall incident:

a. Review Common Responsibilities (Page 1-2).
b. Obtain briefing from the Hazardous Materials Group Supervisor.
c. Participate in the preparation of, and implement the Site Safety and Control Plan (ICS Form 208).
d. Advise the Hazardous Materials Group Supervisor (or Hazardous Materials Branch Director) of deviations from the Site Safety and Control Plan (ICS Form 208) or any dangerous situations.
e. Has authority to alter, suspend, or terminate any activity that may be judged to be unsafe.
f. Ensure the protection of the Hazardous Materials Group personnel from physical, environmental, and chemical hazards/exposures.
g. Ensure the provision of required emergency medical services for assigned personnel and coordinate with the Medical Unit Leader.
h. Ensure that medical related records for the Hazardous Materials Group personnel are maintained.
i. Maintain Unit/Activity Log (ICS Form 214).

TECHNICAL SPECIALIST-HAZARDOUS MATERIALS REFERENCE - Reports to the Hazardous Materials Group Supervisor (or Hazardous Materials Branch Director, if activated). This position provides technical information and assistance to the Hazardous Materials Group using various reference sources such as computer databases, technical journals, CHEMTREC, and phone contact with facility representatives. The Technical Specialist-Hazardous Materials Reference may provide product identification using hazardous categorization tests and/or any other means of identifying unknown materials:

a. Review Common Responsibilities (Page 1-2).
b. Obtain briefing from the Planning Section Chief or assigned supervisor.
c. Provide technical support to the Hazardous Materials Group Supervisor.
d. Maintain communications and coordinate operations with the Entry Leader.
e. Provide and interpret environmental monitoring information.
f. Provide analysis of hazardous material sample.
g. Determine personal protective equipment compatibility to hazardous material.
h. Provide technical information of the incident for documentation.
i. Provide technical information management with public and private agencies i.e.: Poison Control Center, Toxicology Center, CHEMTREC, State Department of Food and Agriculture, National Response Team.
j. Assist Planning Section with projecting the potential environmental effects of the release.
k. Maintain Unit/Activity Log (ICS Form 214).

SAFE REFUGE AREA MANAGER - The Safe Refuge Area Manager reports to the Site Access Control Leader and coordinates with the Decontamination Leader and the Entry Leader. The Safe Refuge Area Manager is responsible for evaluating and prioritizing victims for treatment, collecting information from the victims, and preventing the spread of contamination by these victims. If there is a need for the Safe Refuge Area Manager to enter the Contamination Reduction Zone in order to fulfill assigned responsibilities then the appropriate Personal Protective Equipment shall be worn:

a. Review Common Responsibilities (Page 1-2).
b. Establish the Safe Refuge Area within the Contamination Reduction Zone adjacent to the Contamination Reduction Corridor and the Exclusion Control Line.
c. Monitor the hazardous materials release to ensure that the Safe Refuge Area is not subject to exposure.
d. Assist the Site Access Control Leader by ensuring the victims are evaluated for contamination.
e. Manage the Safe Refuge Area for the holding and evaluation of victims who may have information about the incident, or if suspected of having contamination.
f. Maintain communications with the Entry Leader to coordinate the movement of victims from the Refuge Area(s) in the Exclusion Zone to the Safe Refuge Area.
g. Maintain communications with the Decontamination Leader to coordinate the movement of victims from the Safe Refuge Area into the Contamination Reduction Corridor, if needed.
h. Maintain Unit/Activity Log (ICS Form 214).

ASSISTING AGENCIES

LAW ENFORCEMENT - Local, State, and Federal law enforcement agencies may respond to Hazardous Materials incidents. Depending on incident factors, law enforcement may be a partner in Unified Command or may participate as an assisting agency. Some functional responsibilities that may be handled by law enforcement are:

a. Isolate the incident area
b. Manage crowd control
c. Manage traffic control
d. Manage public protective action
e. Provide scene management for on-highway incidents

f.   Manage criminal investigations

g.   Evidence collection

ENVIRONMENTAL HEALTH AGENCIES - In most cases the local or State environmental health agency will be at the scene as a partner in Unified Command. Some functional responsibilities that may be handled by environmental health agencies are:

a.   Determine the identity and nature of the Hazardous Materials.

b.   Establish the criteria for clean-up and disposal of the Hazardous Materials.

c.   Declare the site safe for re-entry by the public.

d.   Provide the medical history of exposed individuals.

e.   Monitor the environment.

f.   Supervise the clean up of the site.

g.   Enforce various laws and acts.

h.   Determine legal responsibility.

i.   Provide technical advice.

j.   Approve funding for the cleanup.

CIVIL SUPPORT TEAM (CST) – The California National Guard (CNG) Weapons of Mass Destruction Civil Support Teams (CST) are designed to support local incident commanders and local emergency first responders twenty-four (24) hours a day, seven days per week for any Weapons of Mass Destruction (WMD) terrorist event.

CONTROL ZONE LAYOUT

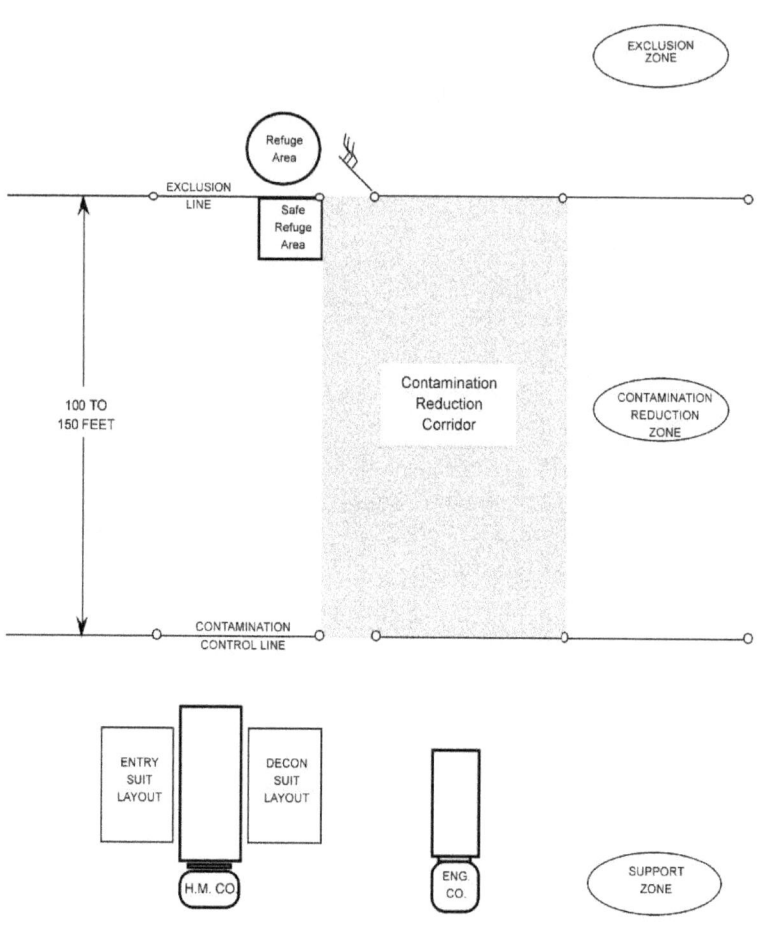

HAZARDOUS MATERIALS COMPANY TYPES
COMPANY TYPING AND MINIMUM STANDARDS

Components	Type 1	Type 2	Type 3
Field Testing	Known Chemicals	Known Chemicals	Known Chemicals
	Unknown Chemicals	Unknown Chemicals	
	WMD Chem / Bio		
Air Monitoring	Combustible Gas Oxygen Carbon Monoxide Hydrogen Sulfide	Combustible Gas Oxygen Carbon Monoxide Hydrogen Sulfide	Combustible Gas Oxygen Carbon Monoxide Hydrogen Sulfide
	Specialty Gases Hydrocarbon Liquid Vapors	Specialty Gases Hydrocarbon Liquid Vapors	
	WMD Chem / Bio		
Sampling: **Capturing** **Labeling** **Evidence** **Collection**	Known Chemicals	Known Chemicals	Known Chemicals
	Unknown Chemicals	Unknown Chemicals	
	WMD Chem / Bio		
Radiation **Monitoring** **And** **Detection**	Gamma	Gamma	Gamma
	Beta	Beta	Beta
	Alpha	Alpha	
	Radio Nuclides		
Chemical **Protective** **Clothing:**	Liquid-Splash Protective	Liquid-Splash Protective	Liquid-Splash Protective
	Vapor Protective	Vapor Protective	
	WMD Chem. / Bio Vapor Protective		
	WMD Chem. / Bio Liquid Splash Protective		

Components	Type 1	Type 2	Type 3
Chemical Protective Clothing: Gloves - Boots	NFPA Compliant Replacement	NFPA Compliant Replacement	NFPA Compliant Replacement
	Hi-Temp. Protective Gloves Cryogenic Protective Gloves	Hi-Temp. Protective Gloves Cryogenic Protective Gloves	
	Radiation Protection Gloves		
Technical Reference	Printed and Electronic	Printed and Electronic	Printed and Electronic
	Plume Air Modeling, Map Overlays	Plume Air Modeling, Map Overlays	
	WMD Chem / Bio Sources		
Special Capabilities	Heat Sensing	Heat Sensing	
	Night Vision	Night Vision	
	Digital Photo	Digital Photo	
	Digital Video		
Intervention	Diking, Damming, Absorption	Diking, Damming, Absorption	Diking, Damming, Absorption
	Liquid, Solid Leak Intervention	Liquid, Solid Leak Intervention	Liquid, Solid Leak Intervention
	Vapor Leak Intervention	Vapor Leak Intervention	
	Neutralization, Plugging, Patching	Neutralization, Plugging, Patching	
	WMD Chem / Bio Spill Containment		

Components	Type 1	Type 2	Type 3
Decontamination	Known Chemicals	Known Chemicals	Known Chemicals
	Unknown Chemicals	Unknown Chemicals	
	WMD Chem / Bio		
Communications	In-Suit	In-Suit	In-Suit
	Cell Phone	Cell Phone	Cell Phone
	Wireless Fax, Copy, Web Access	Wireless Fax, Copy, Web Access	
Respiratory Protection	SCBA	SCBA	SCBA
	APR or PAPR, WMD Chem / Bio Compliant		
Personnel: **Staffing Levels**	Haz Mat Specialist WMD Chem / Bio 7	Haz Mat Specialist 5	Haz Mat Technician 5

Notes

CHAPTER 15

MULTI-CASUALTY

MEDICAL BRANCH

DEFINITION

The Medical Branch structure is designed to provide the Incident Commander with a basic, expandable system to manage a large number of patients during an incident. If incident conditions warrant, Medical Groups may be established under the Medical Branch Director. The degree of implementation will depend upon the complexity of the incident.

MODULAR DEVELOPMENT

A series of examples for the modular development of the Medical Branch within an incident involving mass casualties are included to illustrate one possible method of expanding the incident organization:

Initial Response Organization: The Incident Commander manages initial response resources as well as all Command and General Staff responsibilities. The first arriving resource with the appropriate communications capability should establish communications with the appropriate hospital or other coordinating facility and become the Medical Communications Coordinator. Other first arriving resources would become Triage Personnel.

Reinforced Response Organization: In addition to the initial response, the Incident Commander establishes a Safety Officer, Triage Unit Leader, a Treatment Unit Leader, Patient Transport Unit Leader and Ambulance Coordinator. Also, patient treatment areas are established and staffed.

Multi-Group Response: All positions within the Medical Group are now filled. The Air Operations Branch is shown to illustrate the coordination between the Ambulance Coordinator and the Air Operations Branch. An Extrication Group is established to free entrapped victims.

Multi-Branch Incident Organization: The complete incident organization shows the Multi-Casualty Branch and other Branches. The Multi-Casualty Branch now has multiple Medical Groups (geographically separate) but only one Patient Transportation Group. This is because all patient transportation must be coordinated through one point to avoid overloading hospitals or other medical facilities.

Multi-Casualty –
Initial Response Organization

INCIDENT COMMANDER

Triage
Unit

Engine

Immediate
Treatment

Medical
Communications
Coordinator

Multi-Casualty Initial Response Organization (example): This example depicts the arrival of an Engine Company and ALS Ambulance. These units find conditions warranting a Multi-Casualty response. The Company Officer assumes Incident Command and engine personnel begin the Simple Triage and Rapid Treatment (START) process by triaging victims and, at the same time, assess any additional hazards (fuel spills, unstable vehicles, etc.). A Paramedic from the ambulance becomes Medical Communications Coordinator (Med. Comm.) while the second member (PM or EMT) begins establishing Treatment Areas beginning with the Immediate Treatment Area.

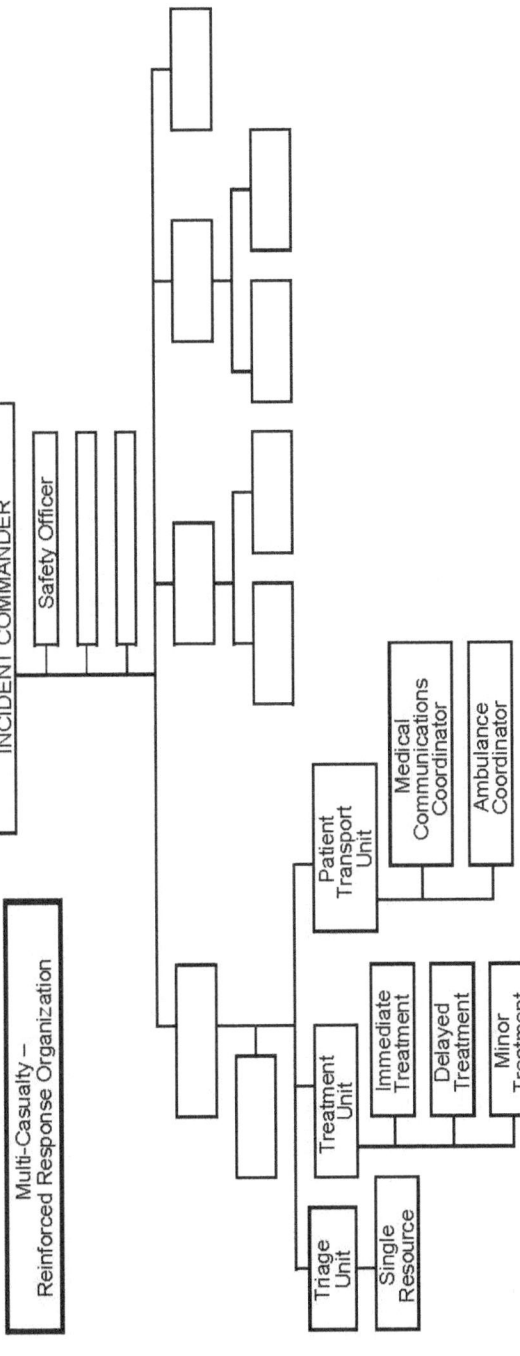

Multi-Casualty – Reinforced Response Organization (example): With the arrival of additional engine companies, an additional ambulance and an Ambulance Supervisor, the Incident Command has established Unit Leaders, reinforced the Treatment Areas, established a Patient Transport Unit and activated an Ambulance Coordinator. A Safety Officer is assigned early in the incident.

Multi-Casualty –
Multi-Group Response Organization

INCIDENT COMMANDER

Safety Officer

Public Information Officer

Operations Section

Staging Area(s)

Medical Group Supervisor

Triage Unit

Single Resource

Morgue Manager

Treatment Unit

Immediate Treatment

Delayed Treatment

Minor Treatment

Treatment Dispatch Manager

Patient Transport Unit

Medical Communications Coordinator

Ambulance Coordinator

Extrication Group Supervisor

Single Resource

Single Resource

Air Operations Branch Director

Resources Unit

Responder Rehabilitation

Multi-Casualty Multi-Group Response Organization (example): The Medical Group Supervisor is managing the treatment and transportation of the injured. In most cases triage would be winding down and those personnel can be assigned to a treatment area. An Air Operations Branch Director is assigned to work with the Patient Transport Unit in coordinating air transportation of patients to distant facilities. The Operations Section Chief has now turned attention to those victims who may be entangled or entombed by establishing an Extrication Group. Other elements of the Command Staff are activated as well as selected elements of the Planning and Logistics Sections.

15-6

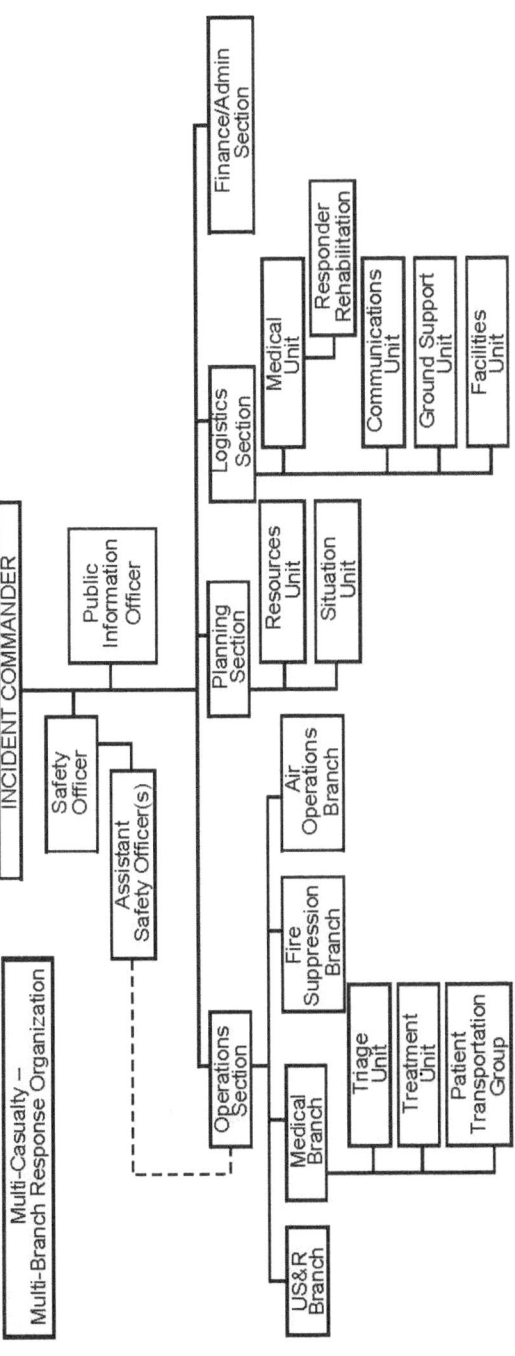

Multi-Casualty – Multi-Branch Response Organization (example): Multiple Medical Groups are working an especially widespread incident. The Patient Transportation Unit has been upgraded to a Group to more effectively handle the multiple transport needs. Other Branches (US&R and Fire Suppression) are activated. Selected Sections and Units of the General Staff are activated. Assistant Safety Officers are assigned within the Operations Section, US&R, and Fire Suppression.

POSITION CHECKLISTS

MEDICAL BRANCH DIRECTOR - The Medical Branch Director is responsible for the implementation of the Incident Action Plan within the Medical Branch. The Branch Director reports to the Operations Section Chief and supervises the Medical Group(s) and the Patient Transportation function (Unit or Group). Patient Transportation may be upgraded from a Unit to a Group based on the size and complexity of the incident:

a. Review Common Responsibilities (Page 1-2).
b. Review Group Assignments for effectiveness of current operations and modify as needed.
c. Provide input to Operations Section Chief for the Incident Action Plan.
d. Supervise Branch activities and confer with Safety Officer to assure safety of all personnel using effective risk analysis and management techniques.
e. Report to Operations Section Chief on Branch activities.
f. Maintain Unit/Activity Log (ICS Form 214).

MEDICAL GROUP SUPERVISOR - The Medical Group Supervisor reports to the Medical Branch Director and supervises the Triage Unit Leader, Treatment Unit Leader, Patient Transportation Unit Leader and Medical Supply Coordinator. The Medical Group Supervisor establishes command and controls the activities within a Medical Group:

a. Review Common Responsibilities (Page 1-2).
b. Participate in Medical Branch/Operations Section planning activities.
c. Establish Medical Group with assigned personnel, request additional personnel and resources sufficient to handle the magnitude of the incident.

d. Designate Unit Leaders and Treatment Area locations as appropriate.
e. Isolate Morgue and Minor Treatment Area from Immediate and Delayed Treatment Areas.
f. Request law enforcement/coroner involvement as needed.
g. Determine amount and types of additional medical resources and supplies needed to handle the magnitude of the incident (medical caches, backboards, litters, and cots).
h. Ensure activation or notification of hospital alert system, local EMS/health agencies.
i. Direct and/or supervise on-scene personnel from agencies such as Coroner's Office, Red Cross, law enforcement, ambulance companies, county health agencies, and hospital volunteers.
j. Request proper security, traffic control, and access for the Medical Group work areas.
k. Direct medically trained personnel to the appropriate Unit Leader.
l. Maintain Unit/Activity Log (ICS Form 214).

TRIAGE UNIT LEADER - The Triage Unit Leader reports to the Medical Group Supervisor and supervises Triage Personnel/Litter Bearers and the Morgue Manager. The Triage Unit Leader assumes responsibility for providing triage management and movement of patients from the triage area. When triage has been completed, the Unit Leader may be reassigned as needed:

a. Review Common Responsibilities (Page 1-2).
b. Review Unit Leader Responsibilities (Page 1-3).
c. Develop organization sufficient to handle assignment.

d. Inform Medical Group Supervisor of resource needs.
e. Implement triage process.
f. Coordinate movement of patients from the Triage Area to the appropriate Treatment Area.
g. Give periodic status reports to Medical Group Supervisor.
h. Maintain security and control of the Triage Area.
i. Establish Morgue.
j. Maintain Unit/Activity Log (ICS Form 214).

TRIAGE PERSONNEL - Triage Personnel report to the Triage Unit Leader and triage patients and assign them to appropriate treatment areas:

a. Review Common Responsibilities (Page 1-2).
b. Report to designated on-scene triage location.
c. Triage and tag injured patients. Classify patients while noting injuries and vital signs if taken.
d. Direct movement of patients to proper Treatment Areas.
e. Provide appropriate medical treatment to patients prior to movement as incident conditions dictate.

MORGUE MANAGER - The Morgue Manager reports to the Triage Unit Leader and assumes responsibility for Morgue Area functions until properly relieved:

a. Review Common Responsibilities (Page 1-2).
b. Assess resource/supply needs and order as needed.
c. Coordinate all Morgue Area activities.
d. Keep area off limits to all but authorized personnel.
e. Coordinate with law enforcement and assist the Coroner or Medical Examiner representative.
f. Keep identity of deceased persons confidential.
g. Maintain appropriate records.

TREATMENT UNIT LEADER - The Treatment Unit Leader reports to the Medical Group Supervisor and supervises Treatment Managers and the Treatment Dispatch Manager. The Treatment Unit Leader assumes responsibility for treatment, preparation for transport, and directs movement of patients to loading location(s):

a. Review Common Responsibilities (Page 1-2).
b. Review Unit Leader Responsibilities (Page 1-3).
c. Develop organization sufficient to handle assignment.
d. Direct and supervise Treatment Dispatch, Immediate, Delayed, and Minor Treatment Areas.
e. Coordinate movement of patients from Triage Area to Treatment Areas with Triage Unit Leader.
f. Request sufficient medical caches and supplies as necessary.
g. Establish communications and coordination with Patient Transportation Unit Leader.
h. Ensure continual triage of patients throughout Treatment Areas.
i. Direct movement of patients to ambulance loading area(s).
j. Give periodic status reports to Medical Group Supervisor.
k. Maintain Unit/Activity Log (ICS Form 214).

TREATMENT DISPATCH MANAGER - The Treatment Dispatch Manager reports to the Treatment Unit Leader and is responsible for coordinating with the Patient Transportation Unit Leader (or Group Supervisor if established), the transportation of patients out of the Treatment Areas:

a. Review Common Responsibilities (Page 1-2).

b. Establish communications with the Immediate, Delayed, and Minor Treatment Managers.
c. Establish communications with the Patient Transportation Unit Leader.
d. Verify that patients are prioritized for transportation.
e. Advise Medical Communications Coordinator of patient readiness and priority for transport.
f. Coordinate transportation of patients with Medical Communications Coordinator.
g. Assure that appropriate patient tracking information is recorded.
h. Coordinate ambulance loading with the Treatment Managers and ambulance personnel.
i. Maintain Unit/Activity Log (ICS Form 214).

IMMEDIATE TREATMENT AREA MANAGER - The Immediate Treatment Area Manager reports to the Treatment Unit Leader and is responsible for treatment and re-triage of patients assigned to Immediate Treatment Area:

a. Review Common Responsibilities (Page 1-2).
b. Request or establish Medical Teams as necessary.
c. Assign treatment personnel to patients received in the Immediate Treatment Area.
d. Ensure treatment of patients triaged to the Immediate Treatment Area.
e. Assure that patients are prioritized for transportation.
f. Coordinate transportation of patients with Treatment Dispatch Manager.
g. Notify Treatment Dispatch Manager of patient readiness and priority for transportation.
h. Assure that appropriate patient information is recorded.
i. Maintain Unit/Activity Log (ICS Form 214).

DELAYED TREATMENT AREA MANAGER - The Delayed Treatment Area Manager reports to the Treatment Unit Leader and is responsible for treatment and re-triage of patients assigned to Delayed Treatment Area:

a. Review Common Responsibilities (Page 1-2).
b. Request or establish Medical Teams as necessary.
c. Assign treatment personnel to patients received in the Delayed Treatment Area.
d. Ensure treatment of patients triaged to the Delayed Treatment Area.
e. Assure that patients are prioritized for transportation.
f. Coordinate transportation of patients with Treatment Dispatch Manager.
g. Notify Treatment Dispatch Manager of patient readiness and priority for transportation.
h. Assure that appropriate patient information is recorded.
i. Maintain Unit/Activity Log (ICS Form 214).

MINOR TREATMENT AREA MANAGER - The Minor Treatment Area Manager reports to the Treatment Unit Leader and is responsible for treatment and re-triage of patients assigned to Minor Treatment Area:

a. Review Common Responsibilities (Page 1-2).
b. Request or establish Medical Teams as necessary.
c. Assign treatment personnel to patients received in the Minor Treatment Area.
d. Ensure treatment of patients triaged to the Minor Treatment Area.
e. Assure that patients are prioritized for transportation.

f. Coordinate transportation of patients with Treatment Dispatch Manager.
g. Notify Treatment Dispatch Manager of patient readiness and priority for transportation.
h. Assure that appropriate patient information is recorded.
i. Maintain Unit/Activity Log (ICS Form 214).

PATIENT TRANSPORTATION UNIT LEADER OR GROUP SUPERVISOR - The Patient Transportation Unit Leader reports to the Medical Group Supervisor and supervises the Medical Communications Coordinator, and the Ambulance Coordinator. The Patient Transportation Unit Leader is responsible for the coordination of patient transportation and maintenance of records relating to the patient's identification, condition, and destination. The Patient Transportation function may be initially established as a Unit and upgraded to a Group based on incident size or complexity:

a. Review Common Responsibilities (Page 1-2).
b. Insure the establishment of communications with hospital(s).
c. Designate Ambulance Staging Area(s).
d. Direct the off-incident transportation of patients as determined by The Medical Communications Coordinator.
e. Assure that patient information and destination are recorded.
f. Establish communications with Ambulance Coordinator.
g. Request additional ambulances as required.
h. Notify Ambulance Coordinator of ambulance requests.
i. Coordinate requests for air ambulance transportation through the Air Operations Branch Director.

j.   Coordinate the establishment of the Air Ambulance Helispots with the Medical Branch Director and Air Operations Branch Director.

k.   Maintain Unit/Activity Log (ICS Form 214).

MEDICAL COMMUNICATIONS COORDINATOR - The Medical Communications Coordinator reports to the Patient Transportation Unit Leader, and maintains communications with the hospital alert system to maintain status of available hospital beds to assure proper patient transportation. The Medical Communications Coordinator assures proper patient transportation and destination:

a.   Review Common Responsibilities (Page 1-2).

b.   Establish communications with the hospital alert system.

c.   Determine and maintain current status of hospital/ medical facility availability and capability.

d.   Receive basic patient information and condition from Treatment Dispatch Manager.

e.   Coordinate patient destination with the hospital alert system.

f.   Communicate patient transportation needs to Ambulance Coordinators based upon requests from Treatment Dispatch Manager.

g.   Communicate patient air ambulance transportation needs to the Air Operations Branch Director based on requests from the treatment area managers or Treatment Dispatch Manager.

h.   Maintain appropriate records and Unit/Activity Log (ICS Form 214).

AMBULANCE COORDINATOR - The Ambulance Coordinator reports to the Patient Transportation Unit Leader, manages the Ambulance Staging Area(s), and dispatches ambulances as requested:

a.   Review Common Responsibilities (Page 1-2).
b.   Establish appropriate staging area for ambulances.
c.   Establish routes of travel for ambulances for incident operations.
d.   Establish and maintain communications with the Air Operations Branch Director regarding Air Ambulance Transportation assignments.
e.   Establish and maintain communications with the Medical Communications Coordinator and Treatment Dispatch Manager.
f.   Provide ambulances upon request from the Medical Communications Coordinator.
g.   Assure that necessary equipment is available in the ambulance for patient needs during transportation.
h.   Establish contact with ambulance providers at the scene.
i.   Request additional transportation resources as appropriate.
j.   Provide an inventory of medical supplies available at ambulance Staging Area for use at the scene.
k.   Maintain records as required and Unit/Activity Log (ICS Form 214).

MEDICAL SUPPLY COORDINATOR -The Medical Supply Coordinator reports to the Medical Group Supervisor and acquires and maintains control of appropriate medical equipment and supplies from units assigned to the Medical Group:

a.   Review Common Responsibilities (Page 1-2).

b. Acquire, distribute and maintain status of medical equipment and supplies within the Medical Group.*

c. Request additional medical supplies.*

d. Distribute medical supplies to Treatment and Triage Units.

e. Maintain Unit/Activity Log (ICS Form 214).

* If the Logistics Section were established, this position would coordinate with the Logistics Section Chief or Supply Unit Leader.

SIMPLE TRIAGE AND RAPID TREATMENT (START) SYSTEM FLOWCHART

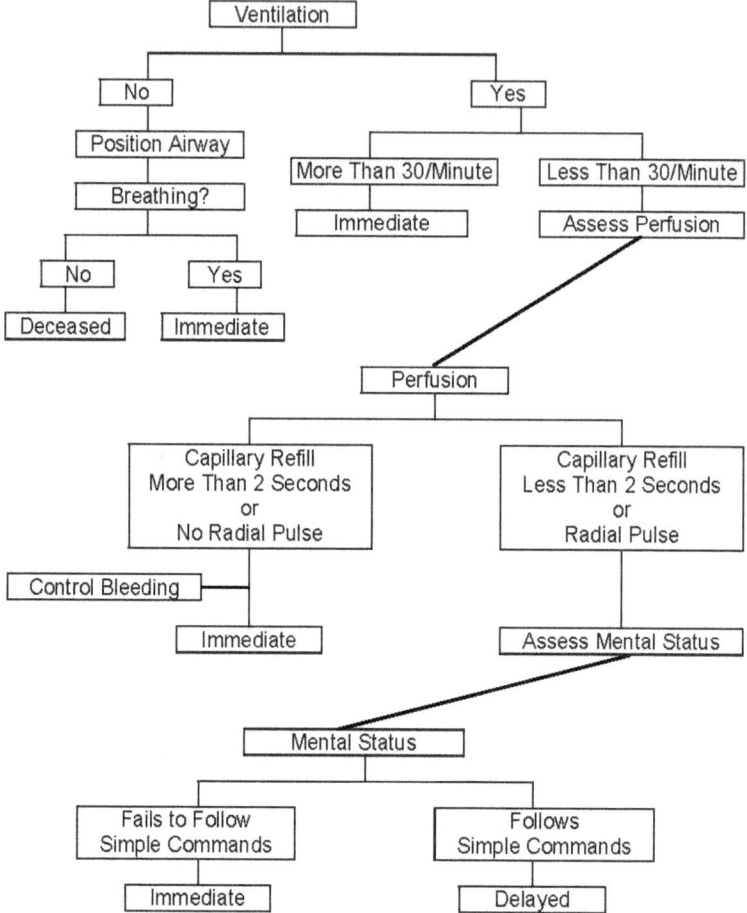

NOTE: Once a patient reaches a triage level indicator in the algorithm (i.e., IMMEDIATE TAG box), triage of this patient should stop and the patient should be tagged accordingly.

CHAPTER 16

URBAN SEARCH AND RESCUE

INTRODUCTION

The Urban Search and Rescue (US&R) organizational module is designed to provide supervision and control of essential functions at incidents where technical rescue expertise and equipment are required for safe and effective rescue operations. US&R incidents can be caused by a variety of events such as an earthquake or terrorist incident that cause widespread damage to a variety of structures and entrap hundreds of people. Other examples of US&R events can range from mass transportation accidents with multiple victims to single site events such as a trench cave-in or confined space rescue involving only one or two victims. US&R operations are unique in that specialized training and equipment are required to mitigate the incident in the safest and most efficient manner possible.

Initial US&R operations will be directed by the first arriving public safety officer who will assume command as the Incident Commander. Subsequent changes in the incident command structure will be based on the resource and management needs of the incident following established ICS procedures.

Additional resources may include US&R Companies and US&R Crews specifically trained and equipped for urban search and rescue operations. The US&R Company is capable of conducting search and rescue operations at incidents where technical expertise and equipment are required. US&R Crews are trained urban search and rescue personnel dispatched to the incident without rescue equipment. US&R Companies and Crews can be assigned as a single resource, grouped to form US&R Strike Teams or added to other resources to form a Task Force. US&R Single Resources, Strike Teams, and Task Forces are managed the same as other incident resources.

Due to the unique hazards and complexity of urban search and rescue incidents, the Incident Commander may need to request a wide variety and amount of multi-disciplinary resources.

US&R Companies and Crews are "typed" based on an identified operational capability. Four levels of US&R operational capability have been identified to assist the Incident Commander in requesting appropriate resources for the incident. These levels are based on five general construction categories and an increasing capability of conducting a rescue at specified emergency situations with an identified minimum amount of training and equipment.

The US&R Type-4 (Basic) Operational Level represents the minimum capability to conduct safe and effective search and rescue operations at incidents involving non-structural entrapment in non-collapsed structures.

The US&R Type-3 (Light) Operational Level represents the minimum capability to conduct safe and effective search and rescue operations at structure collapse incidents involving the collapse or failure of Light Frame Construction and low angle or one-person load rope rescue.

The US&R Type-2 (Medium) Operational Level represents the minimum capability to conduct safe and effective search and rescue operations at structure collapse incidents involving the collapse or failure of Heavy Wall Construction, high angle rope rescue (not including highline systems), confined space rescue (no permit required), and trench and excavation rescue.

The US&R Type-1 (Heavy) Operational Level represents the minimum capability to conduct safe and effective search and rescue operations at structure collapse incidents involving the collapse or failure of Heavy Floor, Pre-cast Concrete and Steel Frame Construction, high angle rope rescue (including highline systems), confined space rescue (permit required), and mass transportation rescue.

The Regional US&R Task Force Level is comprised of 29 people specially trained and equipped for large or complex US&R operations. The multi-disciplinary organization provides five functional elements that include Supervision, Search, Rescue, Medical, and Logistics. The Regional US&R Task Force is totally self-sufficient for the first 24 hours. Transportation and logistical support is provided by the sponsoring agency and may be supported by the requesting agency.

State/National US&R Task Force is comprised of 70 people specially trained and equipped for large or complex US&R operations. The multi-disciplinary organization provides seven functional elements that include Supervision, Search, Rescue, Haz-Mat, Medical, Logistics and Planning. The State/National US&R Task Force is designed to be used as a "single resource." However, each element of the Task Force is modularized into functional components and can be independently requested and utilized.

US&R incidents may occur that will require rescue operations that exceed a resource's identified capability. When the magnitude or type of incident is not commensurate with a capability level, the Incident Commander will have the flexibility to conduct rescue operations in a safe and appropriate manner using existing resources within the scope of their training and equipment until adequate resources can be obtained or the incident is terminated.

ICS MODULAR DEVELOPMENT

The flexibility and modular expansion capabilities of the Incident Command System provides an almost infinite number of ways US&R resources can be arranged and managed. A series of modular development examples are included to illustrate several possible methods of expanding the incident organization based on existing emergency conditions, available resources, and incident objectives.

The ICS Modular Development examples shown are not meant to be restrictive, nor imply these are the only ways to build an ICS organizational structure to manage US&R resources at an incident. To the contrary, the ICS Modular Development examples are provided only to show conceptually how one can arrange and manage resources at an US&R incident that builds from an initial response to a Multi-Branch organization.

ICS MODULAR DEVELOPMENT EXAMPLES

Initial Response Organization (example): The first arriving Public Safety Officer will assume command of the incident as the Incident Commander. The Incident Commander will assume all Command and General Staff functions and responsibilities and manage initial response resources. If the potential for escalation is low, then no specific ICS functional positions are established. If the incident requires an upgraded response, the Incident Commander should consider the early establishment of ICS positions. The following examples illustrate this modular growth of the ICS structure to keep pace with increased resource response.

Reinforced Response Organization (example): In addition to the initial response, more Law Enforcement, local Engine and Truck Companies and Mutual Aid resources have arrived. The Incident Commander forms a Unified Command with the senior ranking Law Enforcement official on scene and has established a Safety Officer to assure personnel safety. A Public Information Officer has been assigned to manage the large media presence. An Operations Section has been assigned to manage the tactical assignments and responsibilities. A Staging Area is established to check in arriving resources. A US&R Group has been established to better coordinate the search and rescue efforts. Public Works is removing debris from the street to improve access and egress routes.

Multi-Group/Division Response Organization (example): The Incident Commander has added a Liaison Officer to the Command Staff to coordinate assisting agencies participation and assigned a Planning and Logistics Section. One US&R Technical Specialist who understands the unique complexities and resource requirements at US&R incidents is assigned to the Planning Section. The Operations Section has established several Groups and Divisions to better coordinate the large volume of diverse resources at the incident. A Law Group and Medical Group have been formed. One State/ National US&R Task Force has arrived and is assigned to Division "A". One Structural Engineer Technical Specialist from the Planning Section is assigned to Division "B" to conduct structural damage assessment. A Handcrew Strike Team is assisting with debris removal.

Multi-Branch Response Organization (example): The Incident Commander has assigned a Finance/ Administration Section. The Operations Section has established five Branches with similar functions to better coordinate and manage resources. The Planning, Logistics and Finance/Administration Section have several Units operational to support the large amount of resources at the incident.

```
Urban Search and Rescue –
Initial Response Organization

                INCIDENT COMMANDER

Engine Company    Truck Company    Engine Company    Ambulance           Law Enforcement
  (Search)          (Rescue)       (Hazard Mitigation) (Medical Treatment) (Perimeter Security)
```

US&R Initial Response Organization (example): The first arriving Public Safety Officer will assume command of the incident as the Incident Commander. The Incident Commander will assume all Command and General Staff functions and responsibilities and manage initial response resources. If the potential for escalation is low, then no specific ICS functional positions are established. If the incident requires an upgraded response, then the Incident Commander should consider the early establishment of ICS positions. The following examples illustrate this modular growth of the ICS structure to keep pace with increased resource response.

Urban Search and Rescue –
Reinforced Response Organization

UNIFIED COMMAND
IC

- Safety Officer
- Public Information Officer

Operations Section
- Staging Area(s)

Medical Group
- Ambulance
- Ambulance
- Engine Company

US&R Group
- Engine Company
- Truck Company
- US&R Company
- US&R Strike Team

Law Enforcement Group
- Law Enforcement
- Law Enforcement
- Traffic Control
- Traffic Control

Public Works
- Debris Removal Equipment

US&R Reinforced Response Organization (example): In addition to the initial response, more Law Enforcement, local Engine and Truck Companies and Mutual Aid resources have arrived. The Incident Commander forms a Unified Command with the senior ranking Law Enforcement official on scene and has established a Safety Officer to assure personnel safety. A Public Information Officer has been assigned to manage the large media presence. An Operations Section has been assigned to manage the tactical assignments and responsibilities. A Stating Area is established to check in arriving resources. A US&R Group has been established to better coordinate the search and rescue efforts. Public Works is removing debris from the street to improve access and egress routes.

Urban Search and Rescue –
Multi-Group Response Organization

UNIFIED COMMAND
IC

Safety Officer
Assistant Safety Officer – US&R
Public Information Officer
Liaison Officer

Operations Section
Logistics Section
Planning Section

Resources Unit
Situation Unit
US&R Technical Specialist

US&R Logistics Technical Specialist

Rapid Intervention Crew/Company
Rapid Intervention Group
Staging Area(s)

US&R Group

Fire Suppression Group
Engine Company
Engine Company
Truck Company

Medical Group
Triage
Treatment
Transportation

Public Works Group (Debris Removal)
Hand Crew(s)
Heavy Equipment

Law Group
Law Enforcement
Traffic Control

Rapid Intervention Crew/Company

Local US&R Company
Mutual Aid US&R Strike Team
Regional US&R Task Force
US&R Structural Technical Specialists

US&R Multi-Group Response Organization (example): The Incident Commander has added a Liaison Officer to the Command Staff to coordinate assisting agencies participation and assigned a Planning and Logistics Section. One US&R Technical Specialist who understands the unique complexities and resource requirements at US&R incidents is assigned to the Planning Section. The Operations Section has established several Groups and Divisions to better coordinate the large volume of diverse resources at the incident. A Law Group and Medical Group have been formed. A Regional US&R Task Force has been assigned to the US&R Group. One State/National US&R Task Force has arrived and is assigned to Division "A." One Structural Engineer Technical Specialist from the Planning Section is assigned to Division "B" to conduct structural damage assessment. A Handcrew Strike Team is assisting with debris removal.

Urban Search and Rescue – Multi-Branch Response Organization

UNIFIED COMMAND
IC

Safety Officer
Assistant Safety Officer(s) (US&R, Haz Mat)
Public Information Officer
Liaison Officer

Operations Section
Planning Section
Logistics Section
Finance/Admin Section

Staging Area(s)

Medical Branch
- Medical Group
- Transportation Group

Haz Mat Branch
- Haz Mat Group
- Decontamination Group

US&R Branch
- US&R Group
- Hazard Control Group
- Fire Suppression Group
- Rapid Intervention Group

Public Works Branch
- Debris Removal Group
- Utilities Group

Law Branch
- Perimeter Control Group
- Traffic Control Group
- Crime Scene Investigation Group
- Force Protection Group

US&R Multi-Branch Response Organization (example): The Incident Commander has assigned a Finance/ Administration Section. The Operations Section has established five Branches with similar functions to better coordinate and manage resources. The Planning, Logistics and Finance/Administration Section have several Units operational to support the large amount of resources at the incident.

16-10

US&R

US&R

POSITION DESCRIPTIONS

ASSISTANT SAFETY OFFICER – URBAN SEARCH AND RESCUE – Reports to the Incident Safety Officer as an Assistant Safety Officer, and coordinates with the appropriate supervisor. The Assistant Safety Officer-US&R must possess the appropriate training to coordinate safety related activities for US&R operations. This position advises the appropriate supervisor on all aspects of health and safety and has the authority to stop or prevent unsafe acts:

a. Review Common Responsibilities (Page 1-2).
b. Obtain briefing from the appropriate supervisor.
c. Participate in the preparation of and implement the incident Site Safety and Control Plan (ICS Form 208).
d. Advise the appropriate supervisor of deviations from the incident Site Safety and Control Plan (ICS Form 208) or any dangerous situations.
e. Has authority to alter, suspend, or terminate any activity that may be judged to be unsafe.
f. Ensure the protection of personnel from physical, environmental, and chemical hazards/exposures.
g. Ensure the provision of required emergency medical services for assigned personnel and coordinate with the Medical Unit Leader.
h. Maintain unit records, including Unit/Activity Log (ICS Form 214).

US&R CANINE SEARCH SPECIALIST – Reports directly to the Search Team Manager. The US&R Canine Search Specialist is responsible for performing the canine search function of the incident. Responsibilities include searching collapsed structures, water, debris piles, land and mudslides, or fire areas as assigned, using appropriate search techniques and dog handler skills. The US&R Canine Search Specialist is responsible for documenting locations of alerts and estimating the status of victims and cooperating with and assisting other search and rescue resources:

a. Review Common Responsibilities (Page 1-2).
b. Obtain briefing from appropriate supervisor.
c. Accountable for all issued equipment.
d. Performs additional tasks or duties as assigned during a mission.
e. Maintain unit records, including Unit/Activity Log (ICS Form 214).

HEAVY EQUIPMENT AND RIGGING SPECIALIST – Initially reports to the Rescue Team Manager and may be assigned where their technical services are required. Responsible for performing construction related liaison to the rescue resources, and for assessing capabilities and the need for various heavy equipment:

a. Review Common Responsibilities (Page 1-2).
b. Participate in the planning of rescue activities.
c. Adhere to all safety procedures.
d. Receive initial briefing from supervisor.
e. Carry out tactical assignments as directed.
f. Conduct an assessment of immediately available cranes and heavy equipment.
g. Inspect equipment condition for safe operation and insure coverage by equipment agreement.
h. Develop a contact list of equipment providers and establish a point of contact.
i. Evaluate and advise on heavy equipment staging area requirements.
j. Brief heavy equipment operators and construction officials regarding rescue operations.

k. Ensure that heavy equipment operators are briefed on rescue site safety considerations and emergency signaling procedures.
l. Identify various rigging techniques to assist in the rescue of victims or stabilization of collapsed buildings, including the development of rigging plans and procedures.
m. Coordinate rigging and heavy equipment utilization for rescue operations with equipment operators and rescue personnel.
n. Keep your immediate supervisor apprised of any tactical accomplishments or conflicts.
o. Participate in operational briefings.
p. Collect and transmit records and logs to Equipment Time Recorder and/or Rescue Team Manager at the end of each operational period.
q. Provide vendor evaluation to Documentation Unit.
r. Maintain unit records, including Unit/Activity Log (ICS Form 214).

US&R TOOL AND EQUIPMENT SPECIALIST – Reports directly to the US&R Task Force Leader. The US&R Tool and Equipment Specialist is responsible for sharpening, servicing and repairing all US&R tools and equipment:

a. Review Common Responsibilities (Page 1-2).
b. Determine personnel requirements.
c. Procure items on site through coordination with Incident Logistics Section.
d. Establish tool inventory and accountability system (appropriate records and reports).
e. Maintain all tools in proper condition.
f. Assemble tools for issuance each operational period per Incident Action Plan.
g. Receive and recondition tools after each operational period.

h. Ensure that all appropriate safety measures are taken in tool conditioning area.
i. Procure equipment during the mobilization phase as directed.
j. Provide accountability and security of the Task Force equipment cache.
k. Maintain unit records, including Unit/Activity Log (ICS Form 214).

US&R MEDICAL SPECIALIST – Reports directly to the US&R Task Force Leader. The Medical Specialist is responsible for providing advanced life support medical care to responders and victims in environments that require special US&R training:

a. Review Common Responsibilities (Page 1-2).
b. Provide emergency medical care to all Task Force personnel and victims in environments requiring specialized US&R training.
c. Develop and implement a Medical Action Plan as specified by the US&R Task Force Leader.
d. Adhere to all safety procedures.
e. Provide accountability, maintenance and minor repairs of assigned medical equipment.
f. Perform additional tasks or duties as assigned during an incident.
g. Maintain unit records, including Unit/Activity Log (ICS Form 214).

RESCUE TEAM MANAGER – Reports directly to the US&R Task Force Leader. Is responsible for managing US&R Rescue Operations and supervising assigned resources:

a. Review Common Responsibilities (Page 1-2).
b. Coordinate, manage, and supervise assigned rescue activities.

c. Adhere to all safety procedures including accountability of personnel.
d. Determine rescue logistical needs.
e. Receive briefings and situation reports and ensuring that all rescue personnel are kept informed of mission objectives and status changes.
f. Provide situation updates and maintain records and reports.
g. Perform additional tasks or duties as assigned during a mission.
h. Provide accountability, maintenance, and minor repairs for all issued equipment.
i. Maintain unit records, including Unit/Activity Log (ICS Form 214).

SEARCH TEAM MANAGER – Reports directly to the US&R Task Force Leader. The Search Team Manager is responsible for managing US&R Search Operations and supervising assigned resources:

a. Review Common Responsibilities (Page 1-2).
b. Develop and implement the tactical search plan.
c. Adhere to all safety procedures including accountability of personnel.
d. Coordinate and supervise all assigned search activities.
e. Determine search logistical needs.
f. Receive briefing and situation reports and ensure that all search personnel are kept informed of status changes.
g. Maintain unit records, including Unit/Activity Log (ICS Form 214).

US&R TECHNICAL SEARCH SPECIALIST – Reports directly to the Search Team Manager. The US&R Technical Search Specialist is responsible for performing the technical search function of the US&R Task Force incident operations:

a. Review Common Responsibilities (Page 1-2).
b. Search areas as assigned using appropriate electronic search equipment and techniques.
c. Document locations of possible finds and if possible, estimate the status of the victim(s).
d. Cooperate with and assist other US&R Resources.
e. Provide accountability for all issued equipment.
f. Perform additional tasks or duties as assigned during an incident.
g. Maintain unit records, including Unit/Activity Log (ICS Form 214).

US&R STRUCTURES SPECIALIST – Reports directly to the Search Team Manager or assigned supervisor. The US&R Structures Specialist is responsible for performing the various structure assessments during incident operations:

a. Review Common Responsibilities (Page 1-2).
b. Assess the structural condition within the area of US&R operations. This includes identification of structure types, specific damage and structural hazards.
c. Recommend the appropriate type and amount of structural hazard mitigation required to minimize the risks to task force personnel.
d. Adhere to all safety procedures.
e. Cooperate with and assist other US&R Resources.
f. Provide accountability, maintenance, and minor repairs for all issued equipment.
g. Perform additional tasks of duties as assigned during an incident.
h. Monitor assigned structures for changes in condition during incident operations.
i. Actively participate in implementation of approved structure hazard mitigation as a designer and/or supervisor.

j.   Coordinate and communicate structure hazard mitigation measures with the Search Team Manager.

k.   Maintain unit records, including Unit/Activity Log (ICS Form 214).

URBAN SEARCH AND RESCUE RESOURCE TYPES

Always use the prefix US&R for Urban Search and Rescue (US&R) Resources.
Order Single Resource or Strike Team by Type (Capability – HEAVY, MEDIUM, LIGHT, or BASIC)

	Type 1 (Heavy)	Type 2 (Medium)	Type 3 (Light)	Type 4 (Basic)
Type	• Heavy Floor Construction • Pre-cast Concrete Construction • Steel Frame Construction • High Angle Rope Rescue (including highline systems) • Confined Space Rescue (permit required) • Mass Transportation Rescue	• Heavy Wall Construction • High Angle Rope Rescue (not including highline systems) • Confined Space Rescue (no permit required) • Trench and Excavation Rescue	• Light Frame Construction • Low Angle Rope Rescue	• Surface Rescue • Non-Structural Entrapment in Non-Collapsed Structures

URBAN SEARCH AND RESCUE RESOURCE TYPES (CONTINUED)

RESOURCE	RADIO	COMPONENT	TYPES			
			1	2	3	4
US&R Company	Agency Identifier USAR (phonetic) Number Identifier (VNC USAR 54)	Equipment Personnel Transportation	Heavy Inventory 6 *	Medium Inventory 6 *	Light Inventory 3 *	Basic Inventory 3 *
US&R Crew **	Agency Identifier Type Identifier Number Identifier (KRN-USAR Crew 2)	Personnel Trained To Appropriate Level Supervision Transportation	6	6	3	3
Regional US&R Task Force	Region Identifier Task Force Number Identifier (R1-TF 1)	Equipment Personnel Transportation	A Regional US&R Task Force is comprised of 29 persons specially trained and equipped for US&R Operations. Personnel from either the Region or Operational Area staff the Regional US&R Task Force.			
State/National US&R Task Force	State ID Task Force Number Identifier (CA-TF 5)	Equipment Personnel Transportation	A State/National US&R Task Force is comprised of 70 persons specially trained and equipped for large or complex US&R Operations. The multi-disciplinary organization provides seven functional elements that include Command, Search, Rescue, Haz-Mat, Medical, Logistics and Plans. These Task Forces are self sufficient for 72 hours.			

* Requests should include vehicle capabilities when necessary (i.e., four wheel drive, off-road truck, etc.)
** The agency/department sending the US&R Crew will identify the Supervisor.

URBAN SEARCH AND RESCUE STRIKE TEAM TYPES AND MINIMUM STANDARDS

Strike Team Types	Number/ Type	Minimum Task Capabilities	Strike Team Leader	Per Single Resource	Total Personnel
Kind US&R					
AR	2 – Type 1 (Heavy)	Vehicle(s) equipped for Heavy Floor Construction, Pre-Cast Concrete Construction, Steel Frame Construction, high angle rope rescue (including highline systems), confined space rescue (permit required), and mass transportation rescue	1	6	13
BR	2- Type 2 (Medium)	Vehicle(s) equipped for Heavy Wall Construction, high angle rope rescue (not including highline systems), confined space (no permit required), and trench and excavation rescue	1	6	13
CR	5 – Type 3 (Light)	Vehicle(s) equipped for Light Frame Construction and low angle rope rescue	1	3	16
DR	5 – Type 4 (Basic)	Vehicle(s) equipped for surface rescue and non-structural entrapment in non-collapsed structure	1	3	16

URBAN SEARCH AND RESCUE STRIKE TEAM TYPES AND MINIMUM STANDARDS (CONT.)

Kind	Strike Team Types	Number/ Type	Minimum Task Capabilities	Strike Team Leader	Per Single Resource	Total Personnel
U S & R	GR	2 – Type 1 (Heavy)	Trained for Heavy Floor Construction, Pre-Cast Concrete Construction, Steel Frame Construction, high angle rope rescue (including highline systems), confined space rescue (permit required), and mass transportation rescue	1	6	13
C R E W	HR	2- Type 2 (Medium)	Trained for Heavy Wall Construction, high angle rope rescue (not including highline systems), confined space (no permit required) and trench and excavation rescue	1	6	13
	IR	5 – Type 3 (Light)	Trained for Light Frame Construction and low angle rope rescue	1	3	16
	JR	5 – Type 4 (Basic)	Trained for surface rescue and non-structural entrapment in non-collapsed structures	1	3	16

US&R SEARCH TEAM TYPES

Search element qualifications and equipment are equivalent on all Canine Types. The differentiating factor is based on the training and certification levels of the canine component. Canine Search Teams will have met all of the capabilities of the preceding types.

RESOURCE	RADIO	COMPONENT	TYPES			
			1	2	3	4
US&R Canine Search Team	Canine Search Team Number Identifier *(Canine Search Team 1)*	Personnel (2) Canine (2) Search Team Manager (1)	• Detections in largest search areas • Detection ability amidst numerous distractions	• Detection in limited sized areas • All general construction categories • Extensive obstacle agility	• Light Frame Construction • Confined areas	• Surface Rescue • Non-structural entrapment in non-collapsed structures • Obstacle agility
US&R Technical Search Team	Technical Search Team Number Identifier *(Tech Search Team 1)*	Personnel (2)	• Audible and optical search equipment to conduct technical search			

TECHNICAL SEARCH TEAM

Kind	Type	Technical Search Strike Team Capability	Strike Team Leader	Technical Search Team	Total Personnel
AT	1	Detection of victims entombed in collapsed or failed structures and environmental mishap with Technical Search equipment	1	2	3

SEARCH TASK FORCE

Resource	Radio Designation	Components	Capabilities	Total Personnel
Search Task Force	Search Task Force	1 – Search Team Manager 1 – Technical Search Team 1 – Canine Search Team	Detection of victims entombed in collapsed or failed structures and environmental mishap with canines and Technical Search equipment.	5

URBAN SEARCH AND RESCUE CANINE SEARCH TEAMS

Search element qualifications and equipment are equivalent on all Canine Types. The differentiating factor is based on the training and certification levels of the canine component. Canine Search Teams will have met all of the capabilities of the preceding types.

Resource	Type 1	Type 2	Type 3	Type 4
US&R Canine	• Detections in largest search areas • Detection ability amidst numerous distractions	• Detection in limited sized areas • All general construction categories • Extensive obstacle agility	• Light Frame Construction • Confined areas	• Surface rescues • Non-structural entrapment in non-collapsed structures • Obstacle agility

OES LAW ENFORCEMENT CANINE RECOVERY TEAMS

Search element qualifications and equipment are equivalent on all Canine Types. The differentiating factor is based on the training and certification levels of the canine component. Canine Search Teams will have met all of the capabilities of the preceding types.

Resource	Type 1 Cadaver Basic	Type 2 Live or Deceased	Type 3 Water
Law Enforcement Canine	• Body above ground • Sub-surface disarticulated • Hanging • Simple structure	• Body above ground • Hanging • Live person, must be area certified • Status of subject unknown	• Submerged • Floating • Shoreline

HEAVY EQUIPMENT RESOURCE TYPING

RESOURCE	COMPONENT	TYPE			
		Type 1	Type 2	Type 3	Type 4
Hydraulic Truck Crane	Rating (Tons) Radius (Feet)	100 ton+ Up to 275 feet	50-100 ton Up to 200 feet	Up to 50 ton Up to 150 feet	
Hydraulic Rough Terrain Crane	Rating (Tons) Radius (Feet)	Up to 50 ton Up to 100 feet			
Conventional Truck Crane	Rating (Tons) Radius (Feet)	150 ton+ Up to 300 feet	75-150 ton Up to 250 feet	Up to 75 ton Up to 150 feet	
Conventional Crawler Crane	Rating (Tons) Radius (Feet)	350 ton+ Up to 350+ feet	100-350 ton Up to 275 feet	Up to 100 ton Up to 160 feet	
Excavator Crawler	Rating (Lbs.) Reach	80k lbs.+ Up to 70 feet	40-80k lbs. Up to 50 feet	Up to 40k lbs. Up to 40 feet	Mini
Loader Rubber Tire	Rating (Cubic Yards)	5 cubic yards	3-5 cubic yards	1-3 cubic yards	Backhoe Skid Steer Mini
Forklift Conventional	Rating (Tons)	25 ton+	10-25 ton	5-10 ton	
Forklift All-Terrain Extendable	Rating (Lbs.)	3-6 tons (6-12k lbs.)			

REGIONAL US&R TASK FORCE

The Regional US&R Task Force Level is comprised of 29 people specially trained and equipped for large or complex US&R operations. The multi-disciplinary organization provides five functional elements that include Supervision, Search, Rescue, Medical, and Tool/Equipment Support. The Regional US&R Task Force is totally self-sufficient for the first 24 hours. Transportation is provided by the sponsoring agency and logistical support will normally be provided by the requesting agency.

A Task Force Leader supervises the Regional US&R Task Force. An Assistant Safety Officer is attached to the Task Force, and upon arrival at the incident, will be supervised by the incident's Safety Officer. The Assistant Safety Officer will work directly with the Task Force Leader and will be assigned to the Task Force's area of operation. The US&R Task Force Search element includes Canine and Technical Search capabilities. The Task Force Rescue element includes a Type 1 US&R Company (personnel and equipment), a Type 1 US&R Crew (personnel), and a Heavy Equipment and Rigging Specialist. This element can conduct rescue operations in all types of structures. The Task Force Medical element is responsible for the care and treatment of injured Task Force members or victims if such care must occur in the hazard area. The Medical element will work within the Incident Medical Unit or directly assigned to the Regional Task Force as appropriate. The tools and equipment support element works within the Task Force for tool and equipment repair and maintenance, and will coordinate with the Incident Logistics Section for acquisition of tools and equipment from off-incident locations.

REGIONAL US&R TASK FORCE ORGANIZATION CHART

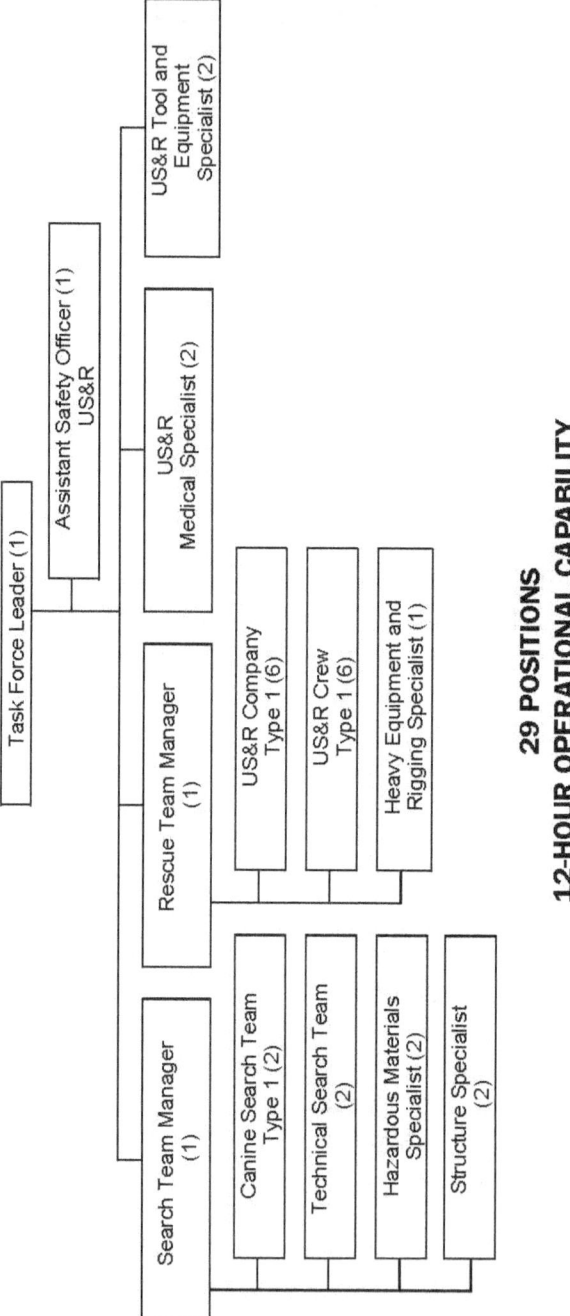

29 POSITIONS
12-HOUR OPERATIONAL CAPABILITY

STATE/NATIONAL US&R TASK FORCE

The Federal Government, through the Federal Emergency Management Agency (FEMA), under the Department of Homeland Security (DHS), has established several State/National Urban Search and Rescue (US&R) Task Forces throughout the nation. All US&R Task Force activities are coordinated through the State Office of Emergency Services (OES) who serves as the primary point of contact for FEMA/DHS. A US&R Task Force is also a State resource that can be acquired without a request for Federal assistance. All requests for a US&R Task Force must go through normal Mutual Aid request procedures. A full, 70-person, Type I, National US&R Task Force is able to deploy within six hours of activation.

Each State/National US&R Task Force is comprised of 70 persons specifically trained and equipped for large or complex US&R Operations. The multi-disciplinary organization provides seven functional elements that include Supervision, Search, Rescue, Haz Mat, Medical, Logistics and Planning. The State/National US&R Task Force can provide round-the-clock US&R Operations (two 12-hour shifts). The US&R Task Force is totally self-sufficient for the first 72 hours and has a full equipment cache to support its operation. Either State or Federal resources provide transportation and logistical support.

A Task Force Leader supervises the State/National US&R Task Force. The US&R Task Force Search element includes physical, canine and electronic capabilities. The Rescue element can conduct rescue operations in all types of structures. The Haz Mat element is primarily responsible for the detection and decontamination of Chemical, Biological, Radiological, Nuclear and Explosive (CBRNE) substances for Task Force members and entrapped victims. The Medical element is primarily responsible for the care and treatment of Task Force members and entrapped victims during extrication. The Logistics element provides the Task Force with logistical support and communications. The Planning element provides personnel competent in structural integrity assessments and documentation of Task Force activities.

The State/National US&R Task Force is designed to be used as a Single Resource, but is modularized into functional elements that can be independently requested and utilized. However, once mobilized as a State/National US&R Task Force, the elements shall remain under the supervision of the US&R Task Force Leader.

A Federal US&R Incident Support Team (IST) coordinates the arrival of a State/National US&R Task Force. The IST is capable of providing overhead management and logistical support to the US&R Task Force while on deployment if an ICS organization is not in place. If an ICS organization is in place, the IST will integrate into that organization. State/ National US&R Task Forces will work within the local incident command organization.

STATE/NATIONAL US&R TASK FORCE ORGANIZATION CHART

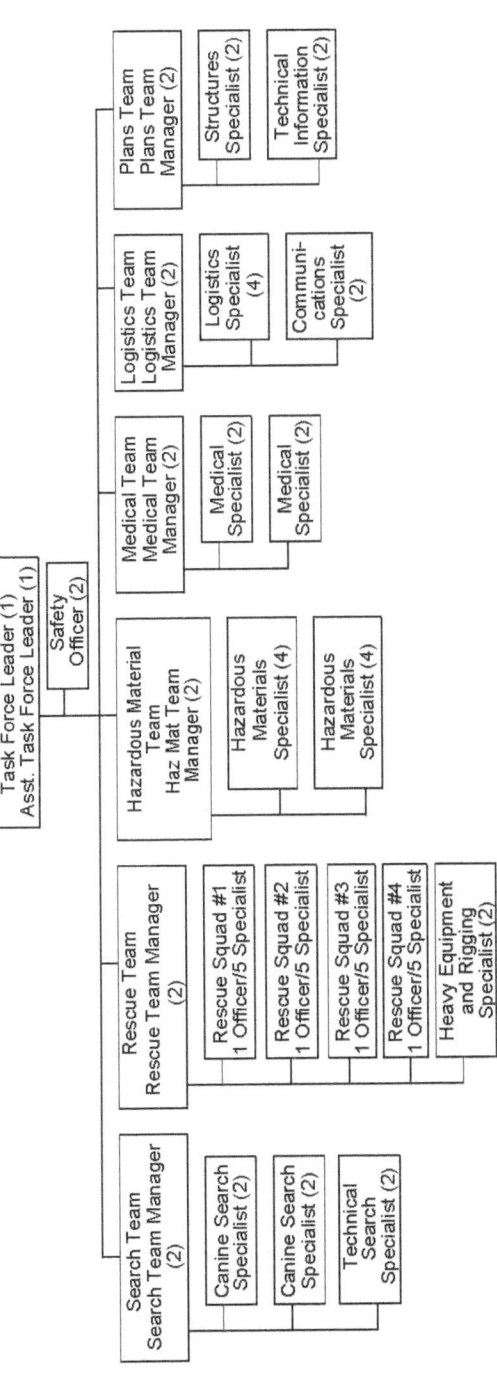

70 POSITIONS

24-HOUR OPERATIONAL CAPABILITY

SELF SUFFICIENT FOR 72-HOURS

US&R

US&R

16-31

STRUCTURE/HAZARDS MARKING SYSTEM

At incidents involving several structures or large areas of damage, the identity and location of individual structures is crucial. The use of existing street names and addresses should always be considered first. If due to damage this is not possible, use the existing hundred block and place all even numbers on one side of the street and all odd numbers on the other side. Mark the new numbers on the front of the structure with orange spray paint. If due to damage the name of the street is not identifiable start with the letter "A" using the phonetic alphabet "Alpha", "Bravo", Charlie, etc.

Structure hazards identified during initial size-up activities and throughout the incident should be noted. This Structure/Hazards Mark should be made on the outside of all normal entry points. Orange spray paint seems to be the most easily seen color on most backgrounds and line marking or downward spray cans apply the best paint marks. Lumber chalk or lumber crayons should be used to mark additional information inside the search mark itself because they are easier to write with than spray paint.

A large square box (approximately two feet) is outlined at any entrance accessible for entry into any compromised structure. Use orange paint for this marking. Specific markings will be clearly made adjacent to the box to indicate the condition of the structure and any hazards found at the time of this assessment. Normally the square box marking would be made immediately adjacent to the entry point identified as safe. An arrow will be placed next to the box indicating the direction of the safe entrance if the Structure/Hazards marking must be made somewhat remote from the safe entrance.

STRUCTURE/HAZARDS MARKINGS

Make a large (2' x 2') square box with orange spray paint on the outside of the main entrance to the structure. Put the date, time, hazardous material conditions and team or company identifier outside the box on the right-hand side. This information can be made with a lumber-marking device.

9/12/93
1310 hrs.
HM – nat. gas
SMA – E-1

Structure is accessible and safe for search and rescue operations. Damage is minor with little danger of further collapse.

9/12/93
1310 hrs.
HM – none
SMA – E-1

Structure is significantly damaged. Some areas are relatively safe, but other areas may need shoring, bracing, or removal of falling and collapse hazards.

9/12/93
1310 hrs.
HM – nat. gas
SMA – E-1

Structure is not safe for search or rescue operations. May be subject to sudden additional collapse. Remote search operations may proceed at significant risk. If rescue operations are undertaken, safe haven areas and rapid evacuation routes should be created.

9/12/93
1310 hrs.
HM – nat. gas
SMA – E-1

Arrow located next to a marking box indicates the direction to a safe entrance into the structure, should the marking box need to be made remote from the indicated entrance.

SEARCH MARKING SYSTEM

Search Markings must be easy to make, easy to read and easy to understand. To be easily seen the search mark must be large and of a contrasting color to the background surface. Orange spray paint seems to be the most easily seen color on most backgrounds and line marking or downward spray cans apply the best paint marks. A lumber marking device may be used to write additional information inside the search mark itself when it would be difficult to write the additional information with spray paint.

A large distinct marking will be made outside the main entrance of each building, structure or area to be searched. This "Main Entrance" search marking will be completed in two steps. First, a large, single slash (approximately two feet) shall be made starting at the upper left moving to the lower right near the main entrance at the start of the search. The Search Team identifier and time that the structure was entered shall be marked to the left of the mid-point of the slash and the date shall be marked near the top of the slash on the opposite side.

When the search of the entire structure is complete and the Search Team exits the building, a second large slash shall be made in the opposite direction forming an "X" on the Main Entrance search marking. Additional information summarizing the entire search of the structure will be placed in three quadrants of the "X". The left quadrant will already contain the Search Team identifier and time when the Search Team first entered the structure. In the top quadrant enter the time the Search Team exited the structure under the date. Change the date if different from date the structure was entered. The right quadrant is for any significant hazards located inside the structure. The bottom quadrant is for the number of live "V" or dead "⩔" victims still inside the structure. Use a small "X" in the bottom quadrant if no victims are inside the structure.

If the search of the entire structure is incomplete, make a circle (approximately one foot in diameter) in the middle of the single slash. The left side will already contain the Search Team identifier and time when the Search Team first entered the structure. At the top end of the slash enter the time the Search Team exited the structure under the date. Change the date if different from date the structure was entered. On the right side, mid-point of the slash, is for any significant hazards located inside the structure. The bottom end of the slash is for the number of live "V" or dead "Ⱶ" victims still inside the structure. Use a small "X" at the bottom if no victims are inside the structure.

During the search function, while inside the structure, a large single slash shall be made upon entry of each room, area or floor. After the search of the room or area has been completed, a second large slash shall be drawn in the opposite direction forming an "X". The only additional information placed in any of the "X" quadrants while inside the structure shall be that pertaining to any significant hazards and the number of live "V" or dead "Ⱶ" victims, as indicated by "V" for live and "Ⱶ" for dead.

SEARCH MARKINGS

Main Entrance Search Marking- WHEN YOU ENTER

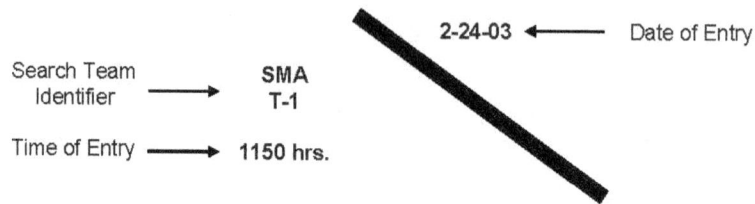

Search Team Identifier → **SMA T-1**

Time of Entry → **1150 hrs.**

2-24-03 ← Date of Entry

Main Entrance Search Marking- WHEN YOU EXIT

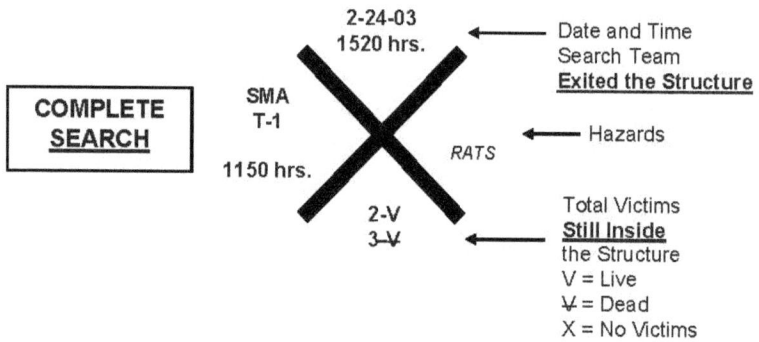

2-24-03
1520 hrs. ← Date and Time Search Team **Exited the Structure**

COMPLETE SEARCH

SMA T-1

1150 hrs.

RATS ← Hazards

2-V
3-V̄ ← Total Victims **Still Inside** the Structure
V = Live
V̄ = Dead
X = No Victims

Main Entrance Search Marking- WHEN YOU EXIT

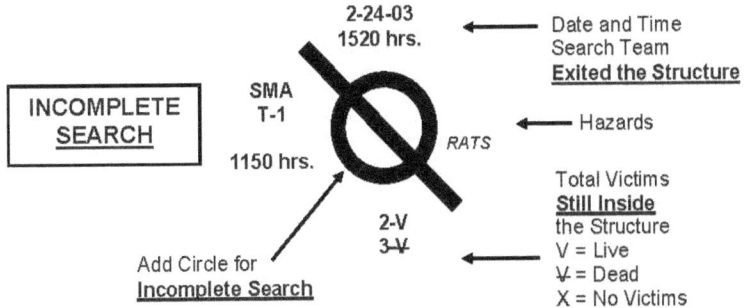

INCOMPLETE
SEARCH

2-24-03
1520 hrs.

SMA
T-1

1150 hrs.

RATS

2-V
3-V

Date and Time
Search Team
Exited the Structure

Hazards

Total Victims
Still Inside
the Structure
V = Live
V = Dead
X = No Victims

Add Circle for
Incomplete Search

Interior Search Markings- EACH ROOM, AREA OR FLOOR

WHEN YOU ENTER

WHEN YOU EXIT

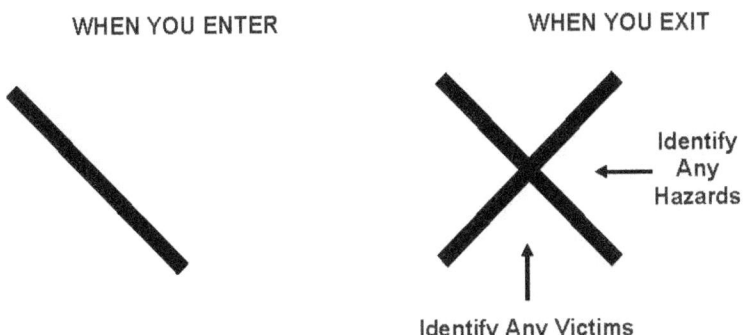

Identify
Any
Hazards

Identify Any Victims

VICTIM MARKING SYSTEM

Make a large (2' x 2') "V" with orange spray paint near the location of a **potential** victim. Mark the name of the Search Team or Crew identifier in the top part of the "V" with paint or a lumber marker type device.

Paint a circle around the "V" when a potential victim is **confirmed** to be **alive** either visually, vocally, or hearing specific sounds that would indicate a high probability of a live victim. If more than one confirmed live victim, mark the total number of victims under the "V".

Paint a horizontal line through the middle of the "V" when a **confirmed** victim is determined to be **deceased**. If more than one confirmed deceased victim, mark the total number of victims under the "V". Use both the live and deceased victim-marking symbols when a combination of live and deceased victims is determined to be in the same location.

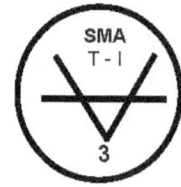

Paint an "X" through the confirmed victim symbol after **all** victim(s) have been removed from the specific location identified by the marking.

An arrow may need to be painted next to the "V" pointing towards the victim when the victim's location is not immediately near where the "V" is painted.

EMERGENCY SIGNALING SYSTEM

Because of the high potential of secondary collapse, dangerous conditions, and the need to communicate other important information, an emergency signaling system should be adopted and in use by all personnel at the incident site. Emergency signals must be a loud and identifiable and sounded when conditions require immediate attention. Emergency signals can be made using devices such as a whistle, air horn, vehicle horn or bell. Each structure or larger area of operations may need to have its own distinct emergency signal device when multiple rescue operations are taking place in the same area to reduce confusion.

Supervisors should identify and inform assigned personnel of a designated place of assembly and/or safe zone for a Personal Accountability Report (PAR) to be conducted should an evacuation signal be sounded. A place of assembly is usually a safe location outside the evacuation area. A safe zone is usually a safe location within a building or disaster site that can be entered within the evacuation area. When an evacuation signal is sounded, all supervisors must conduct a roll call of their assigned personnel and communicate the results of the PAR to their supervisor.

Evacuate the area	**Short signals repeated for 10 seconds, pause for 10 seconds, and repeat for 3 repetitions. Total signal time – 50 seconds.**
Cease Operations/All quiet	**One long signal (8 to 10 seconds).**
Resume Operations	**One long and one short signal.**

Notes

CHAPTER 17

TERRORISM/WEAPONS OF MASS DESTRUCTION (WMD)

INTRODUCTION

Terrorist attacks have created new hazards and responsibilities for the First Responder, whose mission generally includes the protection of life, environment, and property. As first responders continue to be called to emergency incidents (e.g., explosions, hazardous materials (Hazmat) spills, medical responses, fires, etc.), they must now recognize that every incident has the potential of being the result of a terrorist attack. Therefore, first responders must approach each incident aware of the terrorism potential and look for signs that may indicate a terrorist attack in order to take appropriate defensive measures. In general, terrorist attacks will usually present as either hazardous materials and/or USAR (Urban Search and Rescue) events, with a likely multi-casualty result. However, establishing whether an incident is terrorist induced may take authorities hours or even days after the initial danger has passed. Therefore, first responders should refer to their First Responder - Operations (FRO) training for initial actions at such incidents.

Terrorist events manifest from a variety of weaponry: chemical, biological, radiological, nuclear, and explosive (CBRNE). The probability of terrorists using these various devices varies according to their accessibility, transportability and ease of use. Further, it is possible that terrorists could and would use a combination of weapons of mass destruction (WMD) at the same incident. The most common devices used by terrorists are explosives. The most difficult and least likely device that would be used by terrorists is a nuclear device. The least expensive device is one that is chemical or biological in origin.

Recognition clues, warning signs and indicators:

a. Recognition clues may be found in the types of occupancies with a potential risk for a terrorist attack. They may include, but are not limited to:

    1. Government buildings
    2. Mass transit facilities
    3. Public assembly (i.e., sports and entertainment centers)
    4. Places of historic or symbolic value
    5. Religious centers
    6. Family planning centers
    7. Laboratories and testing facilities

b. Warning signs may include, but are not limited to:

    1. Medical incidents of a suspicious nature that produce multiple victims in a non-trauma setting
    2. Explosions in high-risk occupancies
    3. Hazardous materials releases

c. Indicators of possible chemical weapons (CW) usage:

    1. Unusual incidents of dead or dying animals with a lack of insects, or insects on the ground
    2. Unexplained casualties:
        - Multiple victims
        - Serious illnesses
        - Nausea, disorientation, difficulty breathing, convulsions
        - Definite casualty patterns
    3. Unusual liquids, sprays, or vapor:
        - Droplets or oily film
        - Unexplained odors

- Low-lying clouds or fog unrelated to weather
4. Suspicious devices/packages:
  - Unusual metal debris
  - Abandoned spray devices
  - Unexplained munitions

d. Indicators of possible biological weapons (BW) usage:

1. Unusual incidents of sick, dead, or dying animals
2. Unusual casualties:
   - Unusual illness for region or area
   - Definite pattern inconsistent with natural disease
3. Unusual liquids, sprays, or powders
4. Unusual swarms of insects

e. Indicators of suspected radiological or nuclear incident:

1. Simple Radiological Device (SRD), which is a deliberate act of spreading radioactive material without the use of an explosive device (i.e., placement of a radioactive isotopes or radioactive particles on surfaces, air ducts, food, etc.).

2. Radiological Dispersal Device (RDD) or "dirty bomb" which is a combined explosive device with radiological material within. The result is that victims experience an explosion of various magnitudes and are unknowingly contaminated with the resultant radiological material.

3. Improvised Nuclear Device (IND), which is any device designed to cause a nuclear detonation. Construction of such a device is difficult, at best. This is considered a low probability event.

4. Nuclear reactor attacks are considered low probability events due to the high security maintained at these facilities.

DEFINITIONS

Chemical Agents

Terrorists have considered a wide range of toxic chemicals for attacks. Typical plots focus on poisoning foods or spreading the agent on surfaces to poison via skin contact, but may include broader dissemination techniques.

Cyanides

Terrorists have considered using a number of toxic cyanide compounds.

Sodium or potassium cyanides are white-to-pale yellow salts that can be easily used to poison food or drinks. Cyanide salts can be disseminated as a contact poison when mixed with chemicals that enhance skin penetration, but may be detected since most people will notice if they touch wet or greasy surfaces contaminated with the mixture.

Hydrogen cyanide (HCN) and cyanogen chloride (ClCN) are colorless-to-pale yellow liquids that will turn into a gas near room temperature. HCN has a characteristic odor of bitter almonds, and ClCN has an acrid choking odor and causes burning pain in the victim's eyes. These signs may provide enough warning to enable evacuation or ventilation of the attack site before the agent reaches a lethal concentration.

- Both HCN and ClCN need to be released at a high concentration (only practical in an enclosed area) to be effective, therefore leaving the area or ventilating will significantly reduce the agent's lethality.

Exposure to cyanide may produce nausea, vomiting, palpitations, confusion, hyperventilation, anxiety, and vertigo that may progress to agitation, stupor, coma, and death. At high doses, cyanides cause immediate collapse. Medical treatments are available, but they need to be used immediately for severely exposed victims.

Mustard Agent

Mustard is a blister agent that poses a contact and vapor hazard. Its color ranges from clear to dark brown depending on purity, and it has garlic like odor. Mustard is a viscous liquid at room temperature.

- Mustard is not commercially available, but its synthesis does not require significant expertise if a step-by-step procedure with diagrams is available.

Initial skin contact with mustard causes mild skin irritation, which develops into more severe yellow fluid-filled blisters. Inhalation of mustard damages the lungs, causes difficulty breathing, and death by suffocation in severe cases due to water in the lungs. For both skin contact and inhalation, symptoms appear within six to twenty-four hours. There are limited medical treatments available for victims of mustard-agent poisoning.

Nerve Agents

Sarin, Tabun, and VX are highly toxic military agents that disrupt a victim's nervous system by blocking the transmission of nerve signals.

- These agents are not commercially available, and their synthesis requires significant chemical expertise.

Exposure to nerve agents causes pinpoint pupils, salivation, and convulsions that can lead to death. Medical treatments are available, but they need to be used immediately for severely exposed victims.

Toxic Industrial Chemicals

There are wide ranges of toxic industrial chemicals that—while not as toxic as cyanide, mustard, or nerve agents—can be used in much larger quantities to compensate for their lower toxicity.

Chlorine and phosgene are industrial chemicals that are transported in multi-ton shipments by road and rail. Rupturing the container can easily disseminate these gases. The effects of chlorine and phosgene are similar to those of mustard agent.

Organophosphate pesticides such as Parathion are in the same chemical class as nerve agents. Although these pesticides are much less toxic, their effects and medical treatments are the same as for military-grade nerve agents.

Biological Agents

Anthrax

Bacillus anthracis, the bacterium that causes anthrax, is capable of causing mass casualties. Symptoms usually appear within one to six days after exposure and include fever, malaise, fatigue, and shortness of breath. The disease is usually fatal unless antibiotic treatment is started within hours of inhaling anthrax spores; however, it is not contagious.

- Anthrax can be disseminated in an aerosol or used to contaminate food and water.
- Cutaneous anthrax can be caused by skin contact with Bacillus anthracis. This form of the disease, which is easily treated with antibiotics, is rarely fatal.

Botulinum Toxin

Botulinum toxin is produced by the bacterium *Clostridium botulinum*, which occurs naturally in the soil. Crude but viable methods to produce small quantities of this lethal toxin has been found in terrorist training manuals.

- Symptoms usually occur 24 to 36 hours after exposure, but onset of illness may take several days if the toxin is present in low doses. They include vomiting, abdominal pain, muscular weakness, and visual disturbance.
- Botulinum toxin would be effective in small-scale poisonings or aerosol attacks in enclosed spaces, such as movie theaters. The toxin molecule is likely too large to penetrate intact skin.

Ricin

Ricin is a plant toxin that is 30 times more potent than the nerve agent VX by weight and is readily obtainable by extraction from common castor beans. There is no treatment for ricin poisoning after it has entered the bloodstream. Victims show symptoms within hours to days after exposure, depending on the dosage and route of administration.

- Terrorists have looked at delivering ricin in foods and as a contact poison, although we have no scientific data to indicate that ricin can penetrate intact skin.
- Ricin will remain stable in foods as long as they are not heated, and it will have few indicators because it does not have a strong taste and is off-white in color.

Radiological and Nuclear Devices

Radiological Dispersal Devices (RDD)

An RDD is a conventional bomb, not a yield-producing nuclear device. RDD's are designed to disperse radioactive material to cause destruction, contamination, and injury from the radiation produced by the material. An RDD can be almost any size, defined only by the amount of radioactive material and explosives.

- A passive RDD is a system in which unshielded radioactive material is dispersed or placed manually at the target.
- An explosive RDD (often called a "dirty bomb") is any system that uses the explosive force of detonation to disperse radioactive material. A simple explosive RDD consisting of a lead-shielded container (commonly called a "pig") and a kilogram of explosive attached could easily fit into a backpack.

- An atmospheric RDD is any system in which radioactive material is converted into a form that is easily transported by air currents.

Varieties of radioactive materials are commonly available and could be used in an RDD, including Cesium-137, Strontium-90, and Cobalt-60. Hospitals, universities, factories, construction companies, and laboratories are possible sources for these radioactive materials.

Improvised Nuclear Device (IND)

An IND is intended to cause a yield-producing nuclear explosion. An IND could consist of diverted nuclear weapon components, a modified nuclear weapon, or indigenous-designed device.

- IND's can be categorized into two types: implosion and gun assembled. Unlike RDD's that can be made with almost any radioactive material, IND's require fissile material (highly enriched uranium or plutonium) to produce nuclear yield.

PERSONAL SAFETY CONSIDERATIONS

When approaching a scene that may involve chemical, biological, or radiological materials the most critical consideration is the safety of oneself and other responders. Be cognizant that the presence and identification of hazardous agents may not be immediately verifiable, especially in the case of biological and radiological agents. The following actions/measures to be considered by first responders are applicable to either a chemical, biological, or radiological incident. The guidance is general in nature, not all encompassing, and its applicability should be evaluated on a case-by-case basis by the first responders.

Actions to Be Considered:

1. If outside, approach or evacuate upwind of the suspected area.
2. If outside, don available protective mask and clothing immediately. Cover all exposed skin surfaces and protect the respiratory system as much as possible. Personal Protective Equipment (PPE) up to and including self-contained breathing apparatus, and organic vapor respirators will help provide protection.
3. If inside and the incident is inside, evacuate while minimizing passage through the contaminated area, keep windows and doors not used closed.
4. If inside and the incident is outside, stay inside. Turn off air conditioning, seal windows and doors with plastic and tape.
5. If radiological material is suspected, remember to minimize exposure by minimizing time around suspected site, maximizing distance from the site, and trying to place some shielding (e.g. buildings, vehicle, land feature such as a hill, etc.) between yourself and the site.
6. Deploy CBR detection equipment, if available.
7. Report information to the appropriate authorities.

INITIAL RESPONSE

Incident priorities for a suspected or confirmed terrorist incident shall include:

1. Protection of life/health
2. Protection of the environment
3. Protection of the crime scene

4. Protection of property and equipment
5. Preservation of crime scene evidence

At the first indication that an incident may be of a terrorist nature, the first arriving public safety officer shall relocate to a safe location (uphill/upwind/upstream) and institute First Responder Operational (FRO) procedures. Having recognized the extraordinary circumstances of the incident, the first arriving public safety officer may depart from usual FRO procedures and establish an exclusion zone large enough to encompass the number of victims in the affected area and the amount of equipment necessary to accomplish emergency decontamination, plus an area designated as a Safe Refuge Area (SRA).

An artificial line shall be established called the Isolate and Deny Entry (IDE) line to keep unauthorized persons out and to discourage victims from leaving until a more definitive care operation is established; Multi-Casualty, Mass Decontamination (MCMD) or Multi-Casualty Incident (MCI). The first arriving public safety officer shall make the appropriate notifications with respect to the type of CBRNE weapon suspected or discovered at the incident. PPE appropriate for the anticipated hazards of a suspected terrorism related incident should be worn or kept readily available. This may include agency authorized respiratory protection, Mark-1 (atropine, 2-pam/cl) auto-injectors and a personal dosimeter.

The first arriving public safety officer will establish an Incident Command Post that is suitable for the large number of agencies that will respond and participate in this type of incident.

Initial Action Checklist for the First Responder:

Size-up:
   Location by address or intersection
   Incident type: HazMat, MCI, Building Collapse, etc
   Is the incident dynamic or static?
   Is there fire involved?
   How many victims: their signs and symptoms?
   Special instructions: safe approach, Staging, PPE, etc.

Safety:
   PPE
   Weather
   Topography
   Safe distances
   Secondary devices
   Consider all unknown substances lethal until proved
   otherwise

Incident assignments (recommended):
   Establish Unified Command
   Notifications to responsible agencies (local, state and
   federal)
   Determine incident objectives
   Determine an Exclusion Zone
   Establish perimeter control
   Traffic/crowd control
   Emergency decontamination
   Create Safe Refuge Area
   Determine resource needs
   Logistical Support

UNIFIED COMMAND

Unified Command shall be implemented at all Terrorism/WMD incidents when multiple agencies or jurisdictions with statutory or political authority and financial responsibility are involved. Unified Commanders involved in Unified Command shall be collocated. A single Command Post is the best method to ensure effective communications, coordination of resources and overall operational management of the incident.

ICS MODULAR DEVELOPMENT

The flexibility and modular expansion design of the Incident Command System provides a number of ways that public safety and contract resources can be arranged and managed. A series of modular development examples are included to illustrate several possible methods of expanding the incident organization based on existing emergency conditions, available resources and incident objectives.

The ICS Modular Development examples shown are not meant to be restrictive, nor imply these are the only ways to build an ICS organizational structure to manage resources at a Terrorism/WMD incident. To the contrary, the ICS Modular Development examples are provided only to show conceptually how one can arrange and manage resources at the incident that builds from an initial response to a multi-branch organization:

Initial Response Organization (example): The engine company has arrived to find an unknown and suspects that it is a hazardous device. The engine company initiates immediate actions to isolate and evacuate the area. The Company Officer has assumed Incident Command and requested the jurisdictional law enforcement agency to respond and establish Unified Command.

Reinforced Response Organization (example): The potential Terrorism/WMD incident has been reinforced and a Group organization has been created to assist with the management of the incident. Law Enforcement responsibilities of scene security, hazardous device disposal, and crowd and traffic control will be assessed and handled by the appropriate Units/Groups. The Planning Section Chief will accomplish initial planning and resource tracking. The Unified Command will determine the Objectives and the Federal, State and Local agency notification requirements.

Multi-Branch Response Organization (example): As the incident begins to become more complex, the Unified Command decides to create a Law Enforcement Branch and potentially a Fire Branch to address the risks of the incident. Planning and Logistics Sections are partially established to support the resource needs and written action plan. The Unified Command is joined by additional responsible agencies as the incident potential grows.

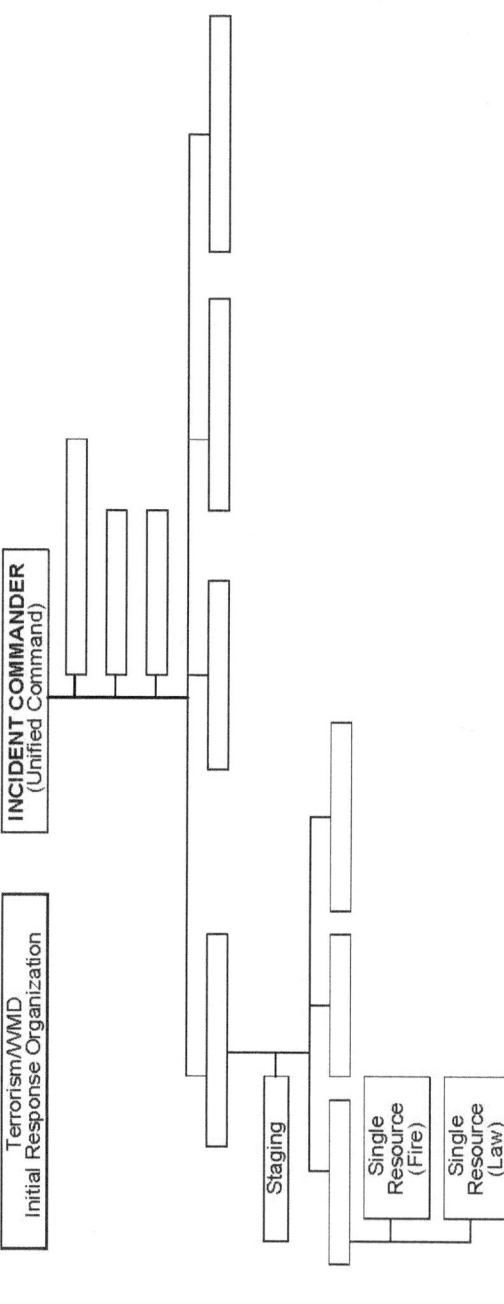

TERRORISM/WMD – Initial Response Organization (example): This chart depicts the initial response organization for a Terrorism/WMD incident.

TERRORISM/WMD

Terrorism/WMD Reinforced Response Organization

INCIDENT COMMANDER
(Unified Command)

- Safety Officer
- Public Information Officer
- Liaison Officer

Operations Section
- Staging
- Divisions
  - Medical Group
    - Single Resource (Fire)
    - Single Resource (Law)
  - Law Group
  - Haz Mat Group
  - Air Operations

Planning Section
- Resources Unit
- Situation Unit
- Intelligence/Investigation Unit

Logistics Section

Terrorism/WMD – Reinforced Response Organization (example): As additional resources arrive, the IC has activated the Operations Section along with multiple Divisions to supervise emergency responder activities. Groups may be assigned certain functions such as medical care for victims, hazardous materials handling or law enforcement activities. Air Operations will coordinate helicopters used for evacuations and reconnaissance. The Planning Section is activated to track and document resource, intelligence and situational status. The Logistics Section is assigned to provide for the service and support needs of the incident.

17-17

TERRORISM/WMD

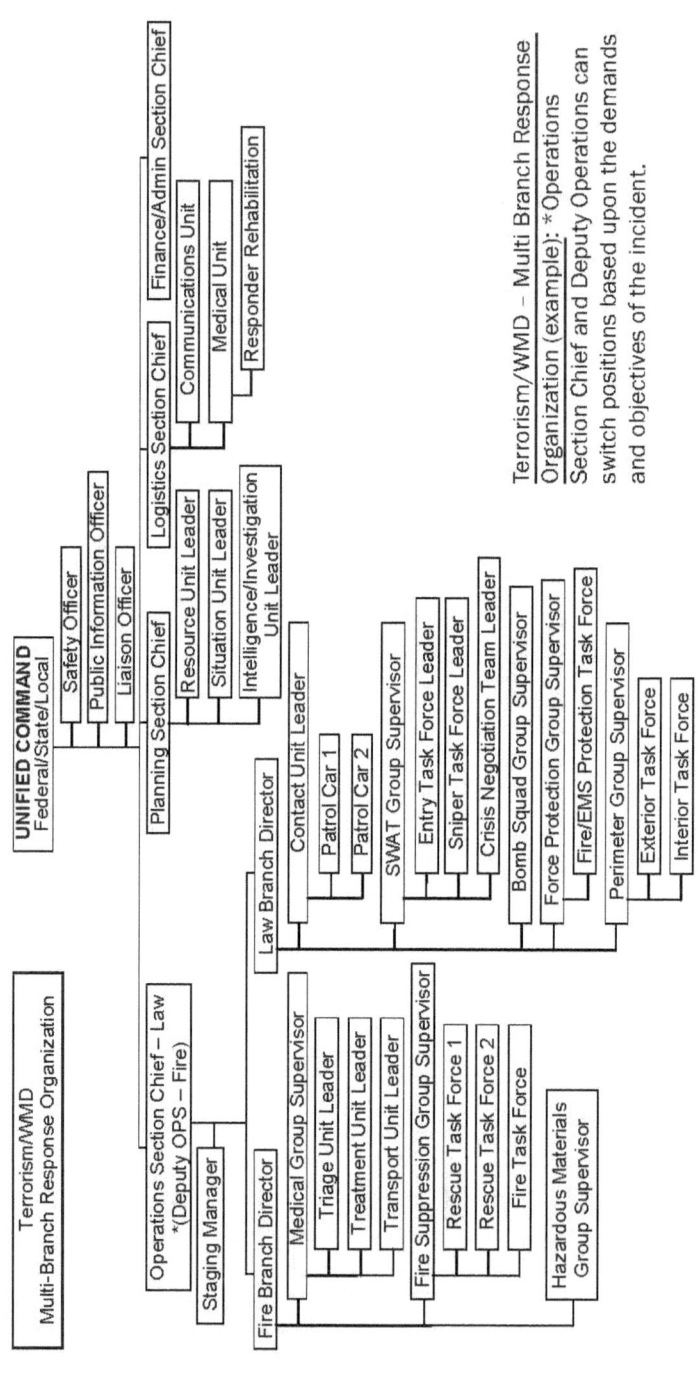

Terrorism/WMD – Multi Branch Response Organization (example): *Operations Section Chief and Deputy Operations can switch positions based upon the demands and objectives of the incident.

POSITION DESCRIPTIONS

INTELLIGENCE UNIT LEADER/GROUP SUPERVISOR –
Initially reports to the Incident Commander, Planning Section Chief or the Operations Section Chief. In a large or complex incident, Intelligence may report to the Law Enforcement Group Supervisor or Branch Director. Based on the needs of the incident, intelligence may be assigned as a Unit under Planning or a Group under Operations/Branch:

a. Coordinates with investigative Unit Leader.
b. Collect and process situational information.
c. Focus on identification of potential suspects.
d. Develop and maintain a working relationship with local, state and federal law enforcement agencies.
e. Obtain, compile and provide intelligence with Operations/ Planning Section Chiefs.
f. Review method of operation by suspect(s).
g. Gather information of suspects and victims.
h. Consider other additional support needs.
i. Maintain Unit/Activity Log (ICS Form 214).

INVESTIGATION UNIT LEADER/GROUP SUPERVISOR –
Initially reports to the Incident Commander, Planning Section Chief or the Operations Section Chief. In a large or complex incident, Investigation may report to the Law Enforcement Group Supervisor or Branch Director. Based on the needs of the incident, Investigation may be assigned as a Unit under Planning or a Group under Operations/Branch:

a. Determine mission and projected length.
b. Determine work location and support requirements.
c. Coordinate with other law enforcement and emergency response agencies.

d. Coordinate intelligence information.
e. Report mission status with the chain of command.
f. Maintain Unit/Activity Log (ICS Form 214).

SECURITY UNIT LEADER/GROUP SUPERVISOR – Initially reports to the Incident Commander, Logistics Section Chief or the Operations Section Chief. In a large or complex incident, Security may report to the Law Enforcement Group Supervisor, Branch Director or Logistics Section Chief. Based on the needs of the incident, Security may be assigned as a Unit under Logistics or a Group under Operations/Logistics or Branch:

a. Determine the security needs of the incident.
b. Determine the scope of the perimeter.
c. Provide incident perimeter and property security.
d. Provide protection for the emergency responders and civilian bystanders.
e. Provide protection to the environment.
f. Control the incident from a safe distance to prevent it from spreading.
g. Facilitate the ingress and egress of emergency resources assigned to the incident.
h. Maintain Unit/Activity Log (ICS Form 214).

HAZARDOUS DEVICE UNIT LEADER/GROUP SUPERVISOR – Initially reports to the Incident Commander or the Operations Section Chief. In a large or complex incident, Hazardous Device may report to the Law Enforcement Group Supervisor, Branch Director or Operations Section Chief. Based on the needs of the incident, Hazardous Device may be assigned as a Unit Leader or a Group Supervisor under Operations/ Branch:

a. Identify the types of hazardous devices at the incident.
b. Determine the location of chemical, biological, radiological, nuclear and explosive devices and to make those devices safe.
c. Determine and communicate the location of safe zones for responders working in the area of hazardous devices.
d. Coordinate with Security to determine the appropriate safe perimeter including fragmentation/inhalation radius.
e. Maintain Unit/Activity Log (ICS Form 214).

Notes

CHAPTER 18

SWIFTWATER/FLOOD SEARCH AND RESCUE

SWIFTWATER/FLOOD SEARCH AND RESCUE
OPERATIONAL SYSTEM DESCRIPTION
ICS US&R 120-2 AND LAW ENFORCEMENT MUTUAL AID
PLAN (SAR) ANNEX

INTRODUCTION

Local and widespread swiftwater and flood emergencies often occur. Many of these incidents strain local resources creating a need for mutual aid resources. This document focuses on the development and identification of specific SF/SAR resources.

This document is intended to provide guidance and develop recommendations for California's SF/SAR resources. This includes but is not limited to:
- Organizational Development
- Resource Typing
- Training and Equipment
- Procedures and Guidelines for Incident Operations

These recommended procedures and guidelines are consistent with both the Standardized Emergency Management System (SEMS) and FIRESCOPE Incident Command System.

It is the responsibility of agencies responding to Mutual Aid, SF/SAR requests, to provide qualified personnel and equipment that meet or exceed the recommended level of skills and capabilities stipulated in this document.

The recommended training, skills, and equipment lists are contained in the Law Enforcement Mutual Aid Plan (SAR) Annex, and the FIRESCOPE Document, ICS-SF-SAR 020-1.

INITIAL RESPONSE

The first arriving public safety officer will direct initial swiftwater/flood search and rescue (SF/SAR) operations. This officer will assume initial command of the operation as the Incident Commander. Subsequent changes in the incident command structure will be based on the needs of the incident, with consideration of jurisdictional responsibilities, established agreements, state and local statutes and shall be accomplished by following established ICS procedures.

Additional resources, specifically trained and equipped for SF/SAR operations may be required. These SF/SAR resources may be assigned as a single resource or grouped together to form Task Forces.

Due to the unique hazards and complexity of SF/SAR incidents, the Incident Commander may require a variety of different multi-disciplinary resources to accomplish the SF/SAR mission (APPENDIX E. Additional SF/SAR Resources).

SF/SAR resources have been categorized or "typed" (APPENDIX A. Swiftwater/Flood Search and Rescue Resource Typing and APPENDIX B. Flood Evacuation Boat Typing). Typing reflects identified operational capabilities, based on specialized training, skills and equipment (ICS SF/SAR 020-1). This typing is based on team qualifications, available equipment and training, as needed for safe and efficient rescue operations for identified SF/SAR tasks.

SF/SAR incidents may occur that will require rescue operations that exceed on-scene personnel capabilities. When the magnitude or type of incident exceeds that capability level, the Incident Commander will have the flexibility to conduct search and rescue operations in a safe and appropriate manner until adequate resources can be obtained or the incident is terminated.

UNIFIED COMMAND

A Unified Command should be implemented at SF/SAR incidents when multiple agencies or jurisdictions with statutory or political authority and financial responsibility are involved. Unified Commanders involved in a Unified Command shall be co-located. A single Command Post is the best method to ensure effective communications, coordination of resources, and overall operational management of the incident.

ICS MODULAR DEVELOPMENT

The flexibility and modular expansion design of the Incident Command System provides an almost infinite number of ways SF/SAR resources can be arranged and managed. Refer to the Law Enforcement Guide for Emergency Operations or the U.S. Fire Administration/National Fire Academy Field Operations Guide (ICS-420-1).

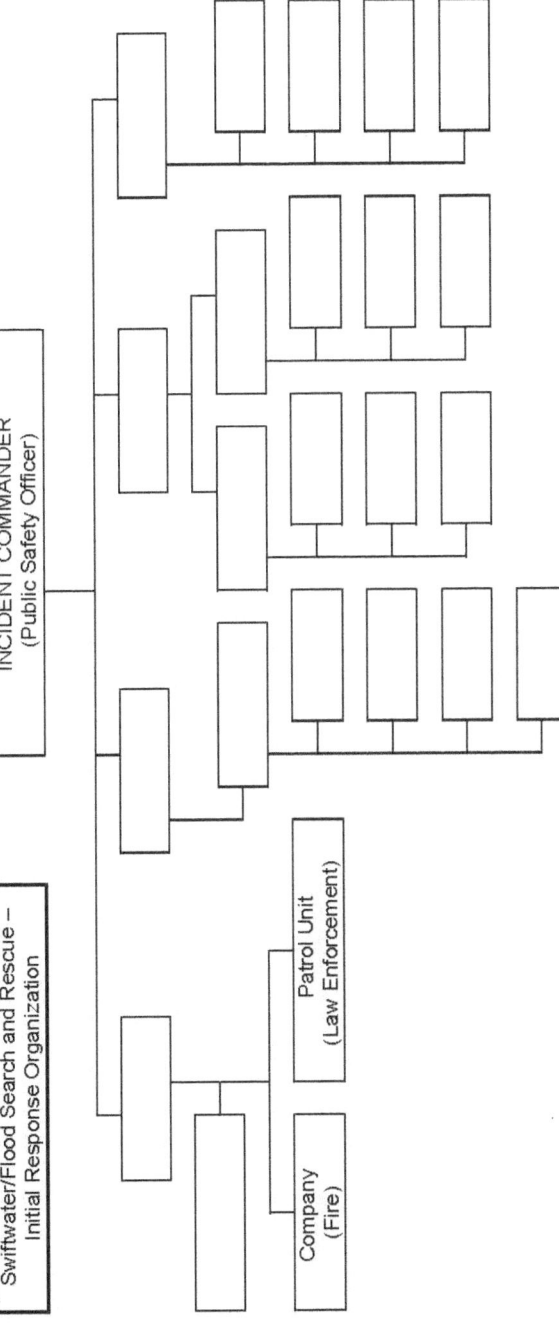

Swiftwater/Flood Search and Rescue – Initial Response Organization (example): The initial Public Safety Officer on scene will assume command of the incident as the Incident Commander. This officer will manage the initial response resources.

Swiftwater/Flood Search and Rescue – Reinforced Response Organization

UNIFIED COMMAND
(Law/Fire/Other Agencies w/jurisdiction)

Safety Officer
Public Information Officer
Liaison Officer

Operations Section

Staging Area(s)

Law Group
- Scene Security (Single Resource)
- Scene Security (Single Resource)
- Traffic Control (Single Resource)
- Traffic Control (Single Resource)

Division A
- Task Force 1 (Law/Fire)
- Engine Strike Team (Fire)
- Ambulance
- SF/SAR Team (Fire)
- SF/SAR Team (Law)

Division B
- Task Force 2 (Law/Fire)
- US&R Strike Team Type 2
- Search (Law)
- Ambulance
- Decontamination Team (Fire)

Swiftwater/Flood Search and Rescue - Reinforced Response Organization (example): Additional Law Enforcement, local Fire Department Engine and Truck Companies, and Mutual Aid resources have arrived. The Incident Commander forms a Unified Command with the designated public safety officials on scene with a Safety Officer, Public Information Officer, and Liaison Officer designated. A Staging Area has been established for arriving resources. The incident is geographically divided into Divisions under an Operations Section. The initial Fire Department resources and/or Law Enforcement SAR Teams are formed into Task Forces. Additional Law Enforcement resources form the Law Group.

SF/SAR

18-6

SF/SAR

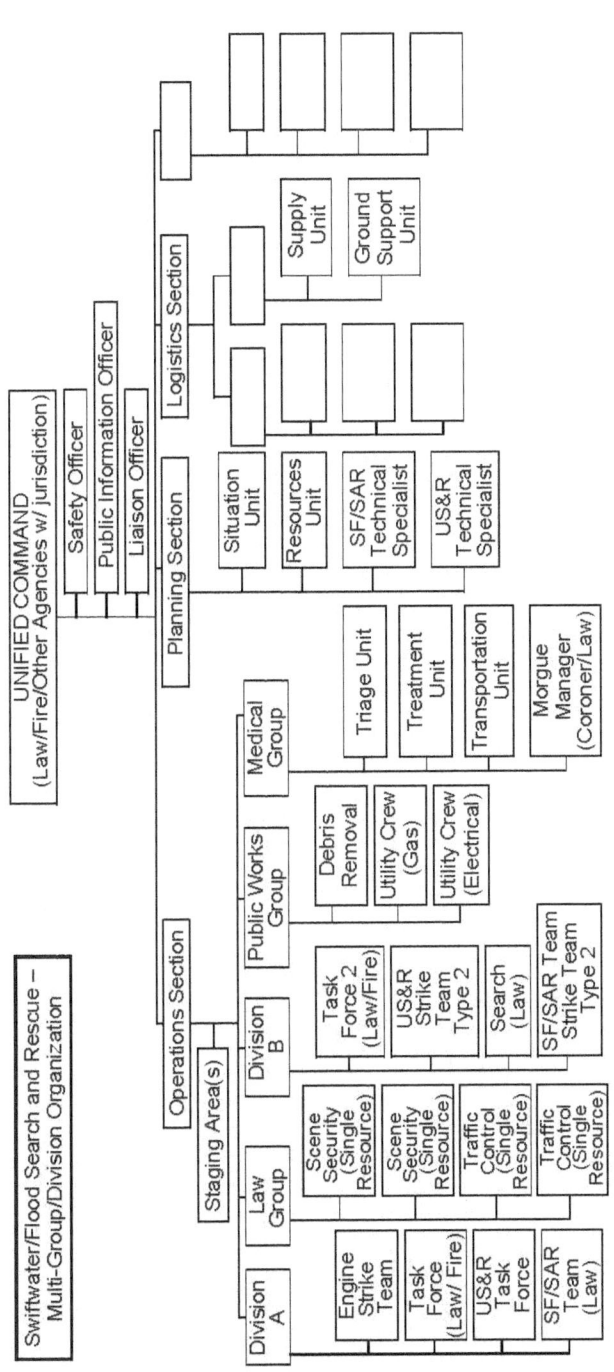

Swiftwater/Flood Search and Rescue – Multi-Group/Division Organization

Swiftwater/Flood Search and Rescue - Multi-Group/Division Organization (example): Planning/Intel and Logistics Sections have been established. Multiple Groups and Divisions have been formed to better manage the incident.

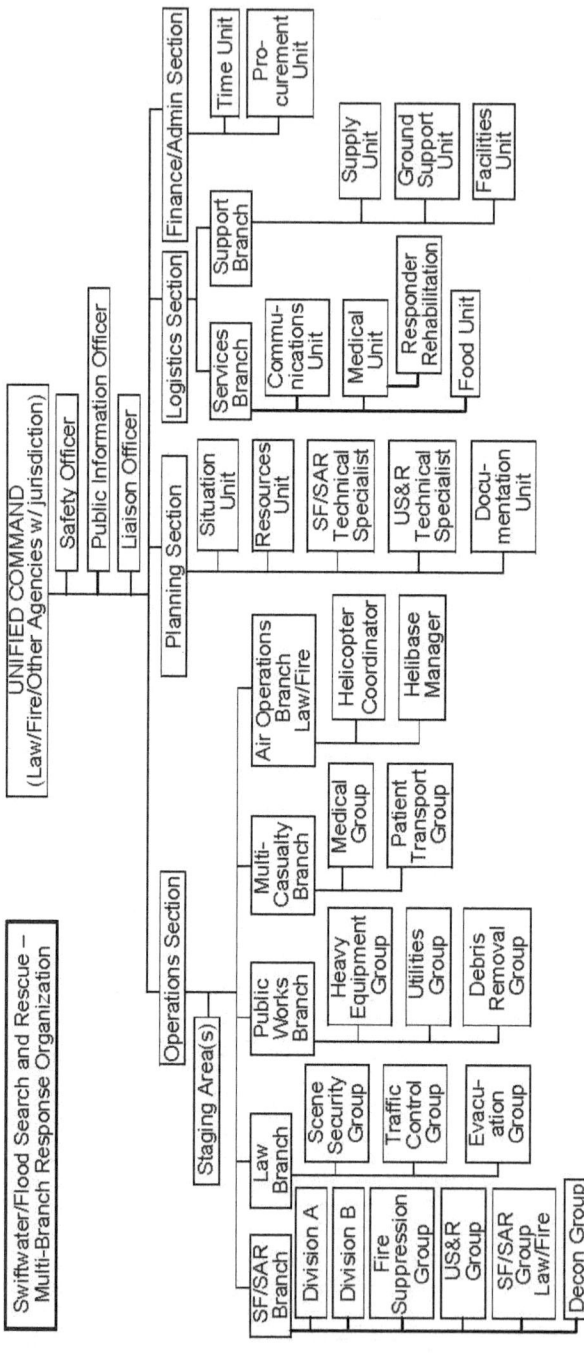

Swiftwater/Flood Search and Rescue - Multi-Branch Response Organization (example): The Incident Commander has assigned Logistics and Finance/Administration Sections.

APPENDIX A - SWIFTWATER/FLOOD SEARCH AND RESCUE RESOURCE TYPING

Type	Type 1	Type 2	Type 3	Type 4
(Capabilities)	Manage search ops Power vessel ops In-water contact rescues Helicopter operational Technical rope systems HazMat Animal rescue EMS-ALS Communications Logistics Capable of 24hr ops	Manage search ops Power vessel ops In-water contact rescues Helicopter operational Technical rope systems HazMat Animal rescue EMS-BLS Capable of 24hr ops	In-water contact rescues Assist in search ops Non-power water craft HazMat Animal rescue EMS-BLS Capable of 24hr ops	Low Risk Land Based HazMat EMS-BLS Helicopter operational

Resource	Component	Type 1	Type 2	Type 3	Type 4
Swiftwater/Flood Search and Rescue Team	Equipment	Type 1 Inventory	Type 2 Inventory	Type 3 Inventory	Type 4 inventory
	Personnel	14 Member Team: *2 Managers* *2 Squad leader* *10 Personnel*	6 Member Team: *1 Squad leader* *5 Personnel*	4 Member Team: *1 Squad leader* *3 Personnel*	3 Member Team: *1 Squad leader* *2 Personnel*
	Transportation	Equipment trailer Personnel transport vehicles	*	*	*

*Requests should include vehicle capabilities when necessary (i.e., four-wheel drive).

APPENDIX B - FLOOD EVACUATION BOAT TYPING

Order these resources by type, quantity, hull design and power type if critical.

Type	Type 1	Type 2	Type 3	Type 4	Type 5
Minimum Victim Transport per Trip	• 5+	• 3 - 5	• 3	• 2	• 2
Special Needs and Notes	• May need launch ramp Power Boat	• May need launch ramp Power Boat	• Hand Launch Power Boat	• Hand Launch • 2 Personal Water Craft (PWC)	• Hand Launch • No Motor • Rafts, skiffs, johnboat, etc.

Resource	Component	Types				
		1	2	3	4	5
Flood Evacuation Boat	Equipment	FEB Inventory	FEB Inventory	FEB Inventory	FEB Inventory	FEB Inventory
	Minimum Personnel	2	2	2	2	2
	Transportation	*	*	*	*	*

*Requests should include vehicle capabilities when necessary (i.e., four-wheel drive).

APPENDIX C - AIR RESOURCE TYPING

Helicopters staffed by personnel trained in search and rescue operations can be ordered through normal Mutual Aid Request procedures. Specify need such as search platform with lights and infrared detectors, hoist capability, swiftwater capability, etc.

Resource	Component	Types			
		1 (Heavy)	2 (Medium)	3 (Light)	4
Helicopter	Seats w/pilot	- 16	- 10	- 5	- 3
	Useful Load (lbs)	- 5000 lbs	- 2500 lbs	- 1200 lbs	- 600 lbs.
	Examples	- UH-60	- Bell 205, 412	- Bell 206, MD 500E, BO 105	- Bell 47 Does not meet mission requirements for external live load.

HELICOPTER Capability/Mission Selection Sheet

*Communications - VHF Programmable Radios

*Over Water Survival Equipment - PFD's for air crew and passengers

□ Live Load *External Load Capable - with rescue equipment
   □ Hoist
   □ Short Haul

□ Sling Load
□ Medical: BLS
□ Medical: ALS
□ Personnel Transportable (number of people)
□ Usable Time (mission duration)
□ Search/Observation

  *Mandatory for aircraft

Mission Equipment Selection Sheet

□ ALS
□ BLS
□ Basket (i.e. Stokes type litter)
□ Cinch Collar
□ Cinch Strap
□ FLIR
□ Night Illumination (1 million candle power +)
□ PA
□ Rescue Capture Ball
□ Rescue Ring
□ Short Haul System
□ Sling Load Capability (in lbs.)
□ Hoist Load Capability (in lbs.)

See next page for Pilot and Flight Crew Capabilities

APPENDIX D - AIR RESOURCE TYPING (PILOT AND CREW)

Pilot Capability

External Load Capable

☐ Victim Location in Static Water
☐ Victim Location in Dynamic Water

- Must be a public service operator, who meets their respective agency's requirement or possesses a USFS, CDF, or OAS (Office of Aircraft Service) valid card.

- Pilot must have a minimum of swiftwater/ flood rescue awareness or operational training along with training and experience in helicopter water rescue evolutions.

Flight Crew Capability

External Load Capable

☐ Victim Location in Static Water
☐ Victim Location in Dynamic Water

- Flight Crew should have a minimum of swiftwater/ flood rescue awareness or operational training along with training and experience in helicopter water rescue evolutions. Aircrew performing water rescue operations must complete annual helicopter water rescue training.

- Areas to include helicopter orientation and safety, hand signals and communications, water rescue device orientation and operations and any additional individual agency specific or type specific requirements.

18-12

APPENDIX E - ADDITIONAL SWIFTWATER/FLOOD
SEARCH AND RESCUE RESOURCES

American Red Cross (ARC). The American Red Cross provides disaster victims assistance such as food, clothing, shelter, and supplemental medical. The ARC provides the emergency mass care to congregate groups and also provides individual/family assistance. Upon the request of government, resources permitting, the ARC may assist with warning, rescue, or evacuations.

Animal Rescue Team. A specialized resource having extensive experience and appropriate equipment required to support the rescue of small domestic pets and large animals' commonly encountered in rural settings. This resource may be available through the Mutual Aid request procedures.

California Conservation Corps (CCC). A State agency that provides personnel for specific non-technical assignments during flood alerts or actual incidents. CCC personnel may be stationed near locations of anticipated problems, due to storm activity, high river tides, or heavy reservoir releases. This resource can be obtained through Mutual Aid request channels.

CAL FIRE (CDF). A State fire agency capable of supplying ICS overhead teams, air assets, fire engines, crews, bulldozers, equipment, camp kitchens, trained personnel for technical or non-technical rescue, containment operations, and storm/flood watch patrols during emergency situations. This resource is available through Mutual Aid request procedures.

U.S. Department of Fish and Wildlife. Federal resources capable of supplying boats with trained operators that include airboats. Orders for specialized equipment must be specific when requesting from this resource through the Mutual Aid request procedure.

Department of Water Resources Flood "Fight" Teams. The Department of Water Resources (DWR) is responsible for coordinating local, state, and federal flood operations. DWR can offer advice to local agencies about how to establish levee patrol, floodwater, place river flood staff gauges, and how to receive flood information from their department. The department can generally assist flood fighting in any area of the state with personnel and flood fighting materials for local agencies. Requests for Flood Fight crews shall be made through the DWR.

Heavy Equipment. Heavy equipment such as cranes, front loaders, and dump trucks are often needed in large quantities during regional water emergencies. They are normally available through local public works departments and private contractors (a pre-signed MOU is recommended). If additional heavy equipment resources are needed, they can be ordered through Mutual Aid request procedure.

Swiftwater/Flood Search and Rescue Technical Specialist. A SF/SAR Technical Specialist may be requested to assist the incident management team with technical expertise in SF/SAR. The specialist is normally assigned to the Planning Section. This resource is ordered through the Mutual Aid request procedure.

Search and Rescue Water Dogs. Dogs specifically scent certified in water, trained to search for and find drowning victims. Search and Rescue Water Dogs are ordered through the Mutual Aid request procedures.

Search Manager. A person qualified and capable of managing the specific search and rescue mission.

Salvation Army. During an emergency, the Salvation Army may be called upon to provide food, clothing, furniture, housing, emergency communication, mobile canteen services, and spiritual ministry for disaster victims. This is generally a local resource, however, it may be requested through the Mutual Aid request procedure.

Structural/Soils Engineers. In most cases, responding resources will have access to local structural and soils engineers through their local agencies. Additional engineers may be ordered through the Mutual Aid request procedure.

Swiftwater/Flood Search and Rescue
Incident Commander Checklist

This list is intended to assist responding public safety personnel with management decisions:

a. Review Common Responsibilities (Page 1-2).
b. Evaluate incident needs.
c. Initiate pre-planned response as appropriate:
   - law enforcement, fire, EMS resources
   - specialized SF/SAR resources
d. Utilize SF/SAR personal protective equipment.
e. Determine additional resource needs.
f. Establish ICS (consider Unified Command).
g. Establish Communication Plan:
   - assign tactical and command channels
   - identify interagency coordination channel(s)
h. Establish resource tracking (personnel accountability) system.
i. Establish search/incident boundaries:
   - identify incident hazards
   - establish operational area
   - manage entry to operational area:
     ◦ limit risk to untrained resources
   - interview reporting party
   - determine victim(s) last known location
j. Consider Evacuation Plan.
k. Consider Traffic Plan/Staging Area(s).
l. Establish down and up stream safety.
m. Implement search and rescue operations:
   - determine rescue vs. recovery
   - evaluate low to high risk options
   - develop contingency plans
n. Establish Medical/Multi-Casualty Plan:
   - consider decontamination of victims
o. Establish logistics support.

SWIFTWATER/FLOOD SEARCH AND RESCUE RECOMMENDED TRAINING, SKILLS AND EQUIPMENT LIST
ICS-SF-SAR 020-1
SF/SAR DECONTAMINATION

Decontamination Of Equipment And Personnel:

The following are the recommended decontamination procedures for resources assigned to SF/SAR operations. Any resources exposed to flood waters during their operations should complete the appropriate level of decontamination. Consult with qualified Hazardous Materials personnel when available.

Basic Decontamination:

Personnel: After completing assignments in floodwaters, hands and face should be washed with clean water and soap. All members should be required to wash hands before entering vehicles and eating areas. Hand washing is essential to reduce secondary contamination.

Equipment: When the team's operational assignment is completed; equipment should be rinsed with clean water. Visible contaminates, mud and light oils, should be removed with soap.

Level 1 Decontamination:

Level 1 decontamination procedures should be used in areas where there is potential for exposure to general contaminates and the water is standing or moving slowly. Examples of areas where the use of this level of decon is needed would be residential and agricultural areas where there is no evidence of large releases of hazardous materials.

Personnel: After completing assignment in floodwaters, hands and face should be washed with clean water and anti-microbial soap (i.e., Vionex or Phisohex). All members should wash their hands before entering vehicles and eating areas. On completion of the day's operations, all members exposed to suspected or known contaminated water should shower and change into clean clothes.

Equipment: When the team's operational assignment is completed, equipment should be washed with soap and clean water. This decon should be completed as soon as possible following the operations. Dry suits should also be washed before entering vehicles for trips from one work site to another.

Level 2 Decontamination:

Level 2 decontamination procedures should be used any time hazardous materials are identified or likely to be present. These include areas of sewage contamination as well as agricultural and chemical contamination. These areas should not be entered, if possible. Limiting the number of personnel exposed to the water should be the top priority of the Team Leader. Support for decontamination should be arranged before units are committed to the contaminated area. **Water samples should be taken for testing from areas entered by the team.** The Medical Unit should be notified if any personnel require this level of decontamination. All personnel exposed to the contaminates should have a one hour, twelve hour, and twenty-four hour medical check following their exposure.

Personnel: After exiting the water, even for short periods during the operational period, members should go through a scrub gross decon* wash with soap and clean water. Remove gloves and wash hands and face with clean water and anti-microbial soap. At the end of the duty period, members should go through a gross decon scrub wash with soap and clean water before any safety gear is removed. Wash hands and face with clean water and anti-microbial soap after removing all safety gear. Shower, using anti-microbial soap before leaving the scene if possible, or as soon as possible thereafter and change into clean clothes.

Equipment: All equipment should be sprayed with bleach solution** or other agents as recommended by on-scene Hazardous Materials personnel and allowed to stand a minimum of fifteen minutes. Thoroughly rinse all treated equipment with clean water and allow to dry before storing with other equipment. Bag any equipment that cannot be dried for the return trip to the base. Wipe with bleach solution** any surfaces inside vehicles that might have come in contact with wet safety equipment during the duty period. Units requiring Level 2 Decontamination should be taken out of service until all equipment has been cleaned and dried.

*   Gross Decon Wash: This is a two-stage process that is set up along a decontamination corridor. All run-off solutions are retained for proper disposal. Persons implementing the corridor should be protected by splash gear. It is recommended that qualified Hazardous Materials personnel be requested to implement this procedure.

Stage 1: Rescuer in safety gear is scrubbed with brushes using a clean water and soap solution. Any contaminated tools are left behind here for cleaning.
Stage 2: Rescuer is rinsed with clean water.

** Bleach Solution: Bleach solution should be made using 30cc of Sodium Hypochlorite 5% (household bleach) for every one gallon of clean water. This will yield a 20,000 ppm solution of bleach.

CHAPTER 19

HIGH-RISE STRUCTURE FIRE INCIDENT

INTRODUCTION

The High-Rise module describes an all-hazard organization designed to provide effective management and control of essential functions at incidents occurring in large, multi-story buildings. These incidents may present significant management, logistical and safety challenges to emergency personnel.

The size and complexity of the interior spaces; limited, sometimes arduous access; with extended travel and response times all contribute to the problems faced by emergency responders.

Additionally, most high-rise structures are equipped with various environmental, fire protection, and life safety systems that require support and control. Successful emergency operations in these types of buildings also require preplanning and technical competence on the part of emergency responders.

MODULAR ORGANIZATION DEVELOPMENT

The order in which the ICS organizational structure develops may vary with the type and scope of the incident. Following are examples of modular development of the ICS that serve to illustrate typical methods of expanding the management organization at a high-rise incident. These examples reflect the size and complexity of the incident and the available resources at a given time in the incident:

Initial Response Organization: The Incident Commander manages the initial response resources as well as all Command and General Staff responsibilities.

Multi-Group/Division Organization: The Incident Commander has established most Command and General Staff positions and has established a combination of divisions and groups to reflect the location and nature of the incident.

Multi-Branch Organization: The Incident Commander has identified a number of actual or potential incident challenges and has established all Command and General Staff positions. The IC has also established several branches to effectively manage the problems and the resources required for mitigation.

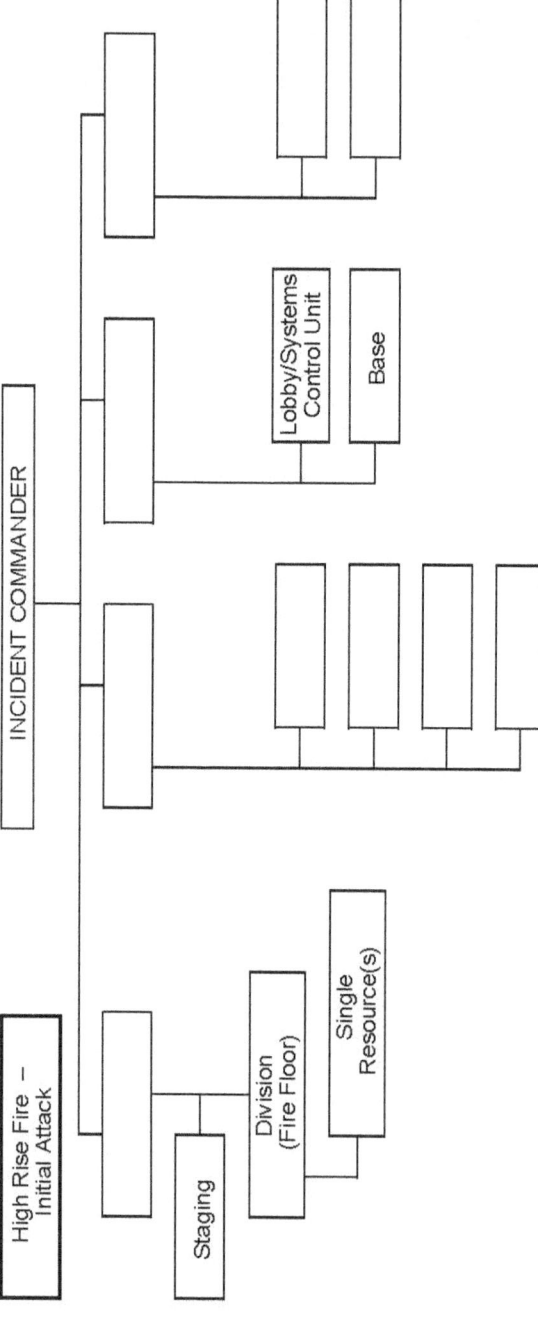

High-Rise Fire Initial Attack (example): This chart depicts the initial assignment including a Command Officer on a fire involving a single floor of a high-rise building. The IC has deployed resources to Fire Attack, Lobby Control, Staging, and Base (ALS-BASE).

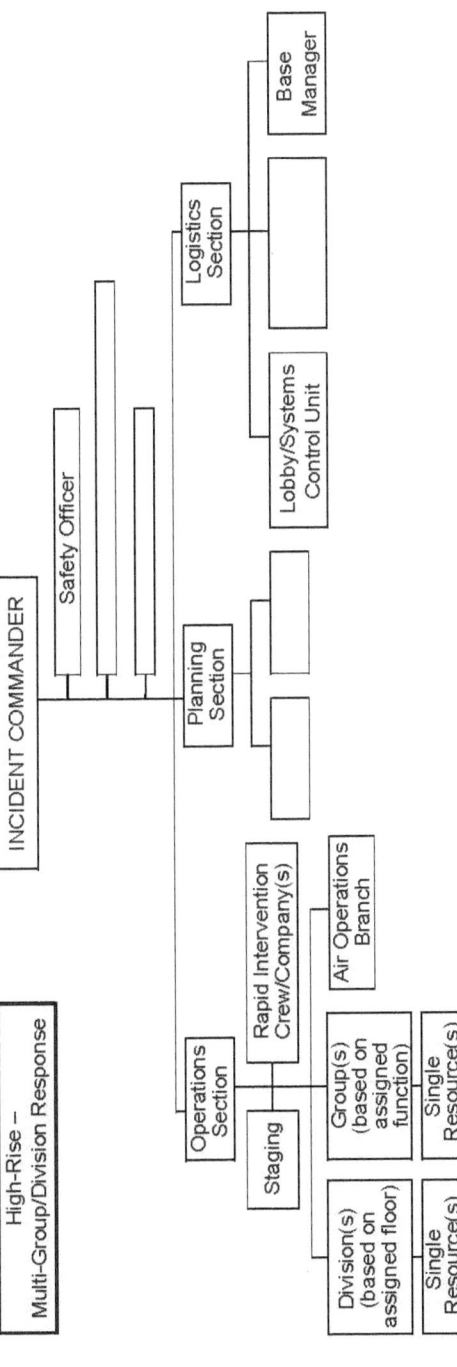

High-Rise Multi-Group/Division Response (example): As additional resources arrive, the IC has activated the Operations Section Chief along with multiple Divisions to supervise action on each involved or threatened floor. Rapid Intervention Crews/ Companies are assigned as determined most effective by Operations. Groups may be assigned certain functions such as medical care for victims, or stairwell pressurization/ventilation. Air Operations Branch will coordinate helicopters used for evacuations or reconnaissance. The Planning Section is activated with selected units. Logistics is assigned to manage Lobby Control, Systems Control, Ground Support, and the Incident Base.

HIGH RISE

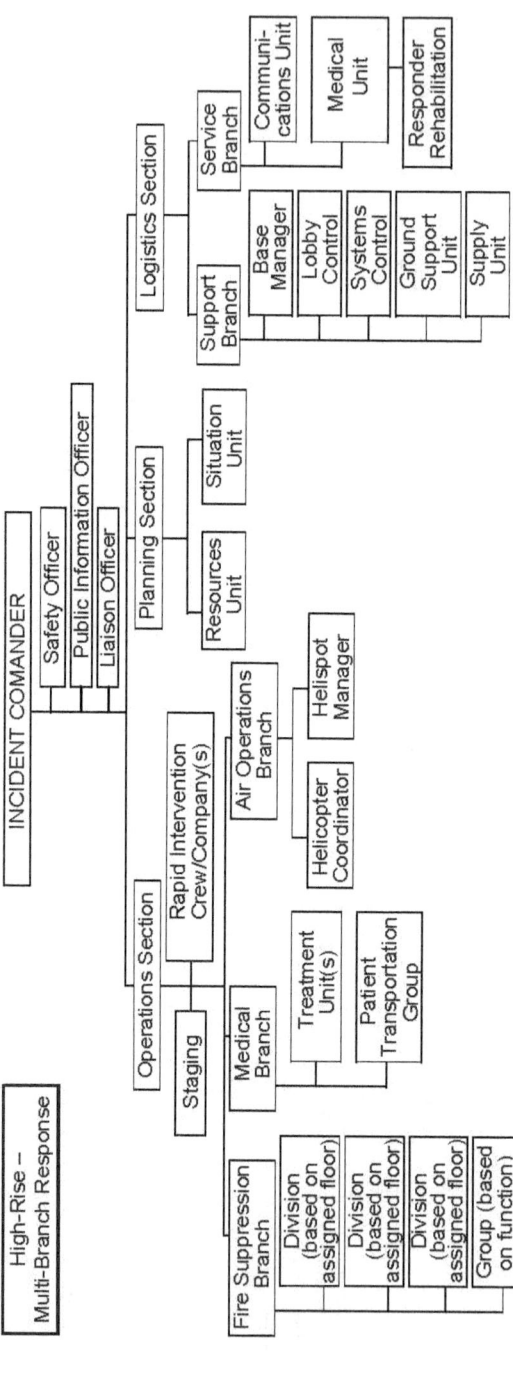

High-Rise Multi-Branch Response (example): The fire has involved multiple floors with various Divisions and Groups assigned. This complexity has led the Operations Section to create a Fire Suppression Branch to manage these Divisions and Groups. A Medical Branch is established and the Air Operations Branch is expanded. The Planning Section has expanded to include the Resources Unit and Situation Unit. Logistics Section has activated the Support and Service Branches as well as various Units within each Branch to accommodate the extensive logistical requirements for this size incident.

DESIGNATED INCIDENT FACILITIES

Base and Staging have modified functions and locations in high-rise incidents:

Staging Area: The challenging nature of high-rise incidents requires modification to the standard ICS concept of a Staging Area. The limited access and vertical travel distance of large high-rise buildings require establishment of a resource Staging Area within the building. The high-rise Staging Area must also serve multiple functions. The Staging Area is generally located a minimum of two floors below the emergency, as long as the atmosphere is tenable. The specific changes are described in the Staging Area Manager's Position Description.

Base: The Base at a high-rise incident resembles a ground level Staging Area. The main difference between Base and a typical Staging Area is that Base must be expanded to perform the functions inherent to supporting large numbers of personnel and equipment. Base should be located away from the incident building to provide for the safety of personnel and equipment.

ORGANIZATION AND OPERATIONS

Modified ICS Positions: Certain existing ICS positions and functional units within the high-rise incident organization have modified responsibilities that require full descriptions. These positions include: Staging Area Manager, Rapid Intervention Group Supervisor, Base Manager, Ground Support Unit Leader, Safety Officer, and Evacuation Group Supervisor.

Specialized High-Rise ICS Positions: Lobby Control and Systems Control Unit Leaders are specialized functional positions specific to a high-rise incident.

Lobby Control Unit is established to provide access control, accountability, and routing inside the building. As the incident escalates, a separate Systems Control Unit may be established to operate, supervise, and coordinate the vital operation of specialized systems incorporated into modern high-rise buildings. These systems may include electrical supply and smoke removal systems. Systems Control Unit coordinates the efforts of various Technical Specialists who might be required to assist in the operation and/or repair of the various systems. During the initial period of an incident, or in a less complex building, the Lobby Control Unit may assume the functions of the Systems Control Unit as shown in the basic organization chart.

The positions and modifications are described in the position checklists that follow. The major responsibilities and procedures for each are further explained in the position manuals.

POSITION CHECKLISTS

HIGH-RISE INCIDENT LOBBY CONTROL UNIT LEADER - The High-Rise Incident Lobby Control Unit Leader's primary responsibilities are as follows: maintain an accountability system, control all building access points and direct personnel to correct routes, control and operate elevator cars, and direct building occupants and exiting personnel to proper ground level safe areas. As directed by the Incident Commander or agency policy, this unit may be assigned the responsibilities of the Systems Control Unit. The Lobby Control Unit Leader reports to the Support Branch Director/ Logistics Section Chief.

The Lobby Control Unit Leader should be prepared to provide the Incident Commander or Planning Section with current information from the personnel accountability process.

The safest method of ascending to upper floors is the use of stairways. The use of elevators for emergency operations should be determined by department policy. This determination is the ultimate responsibility of the Incident Commander; however, the Lobby Control Unit Leader coordinates the actual use of elevators:

a. Check in and obtain briefing from Support Branch Director, Logistics Section Chief or Incident Commander.

b. Make entry, assess situation, and establish Lobby Control position.

c. Request needed resources.

d. Obtain building access keys.

e. Establish entry/exit control at all building access points.

f. Maintain accountability for personnel entering/exiting the building.

g. Assure personnel are directed to the appropriate stairways/elevator for assignment.

h. Control the elevators and provide operators if approved for use by the Incident Commander.

i. Provide briefings and information to Support Branch/ Logistics Section or the Incident Commander.

j. Perform the functions of the Systems Control Unit when directed by the Incident Commander or agency policy.

k. Secure operations and release personnel as determined by the Demobilization Plan.

l. Maintain a Unit/Activity Log (ICS Form 214).

HIGH-RISE INCIDENT SYSTEMS CONTROL UNIT LEADER
The High-Rise Incident Systems Control Unit Leader is responsible for evaluating and monitoring the functions of all

built-in fire protection, life safety, environmental control, communications and elevator systems. The Systems Control Unit Leader may operate, support or augment the systems as required to support the incident plan. The Systems Control Unit Leader reports to the Support Branch Director (if established) or to the Logistics Section Chief. Working with the building's engineering staff, the System Control Unit Leader may respond directly to requests from the Operations Section Chief by using the manual operation modes of the various built-in systems. The Systems Control Unit Leader must establish and maintain a close liaison with building's engineering staff, utility company representatives, and other appropriate technical specialists:

a. Check in and obtain briefing from the Lobby Control Unit, Support Branch Director, Logistic Section Chief or Incident Commander:
   - Briefing must include the type and performance of built-in systems.
   - Introductions to building's engineering staff should occur at briefing.
b. Evaluate current situation and request needed personnel and resources.
c. Establish communication with the building engineer, utility company representatives, elevator service personnel or others to coordinate the operation of selected systems.
d. Assign personnel to monitor all building fire protection/ life safety systems.
e. Evaluate the status and operation of the building's fire and domestic water pumps and water supply (support as needed).
f. Evaluate the operational effectiveness of the heating, ventilation, and air-conditioning system (HVAC); the smoke removal system; and stairwell protection system (support as needed).

g. Evaluate the building's electrical system, emergency power systems, and security systems (support as needed).

h. Evaluate the public address, telephone, emergency phone, and other building communications systems (support as needed).

i. Secure operations and release personnel as determined by the Demobilization Plan.

j. Maintain Unit/Activity Log (ICS Form 214).

HIGH-RISE INCIDENT STAGING AREA MANAGER - The High-Rise Incident Staging Area Manager is responsible for the management of all functions at the Staging Area, and reports to the Operations Section Chief:

a. Obtain briefing from Operations Section Chief or Incident Commander.

b. Proceed to selected location and evaluate suitability:
   • Make recommendations regarding relocation, if appropriate.

c. Request necessary resources and personnel.

d. Establish Staging Area layout and identify/post each functional area i.e., Crew-Ready Area, Air Cylinder Exchange, Equipment Pool, and Medical Unit if collocated within the Staging Area.

e. Determine, establish, or request needed facility services i.e., drinking water and lighting.

f. Coordinate with Logistics Section or Systems Control Unit to maintain fresh air.

g. Maintain a personnel accountability system for arriving and departing crews.

h. Request required resource levels from the Operations Section Chief:
   • Maintain levels and advise the Operations Section Chief when reserve levels are reached.

i. Coordinate with the RIC Group Supervisor to designate area(s) for Rapid Intervention Crew (RIC) to standby if collocated within the Staging Area.
j. Direct crews and equipment to designated locations as requested by the Operations Section Chief or Incident Commander.
k. Secure operations and release personnel as determined by the Demobilization Plan.
l. Maintain Unit/Activity Log (ICS Form 214).

HIGH-RISE INCIDENT RAPID INTERVENTION GROUP SUPERVISOR – The High-Rise Incident Rapid Intervention Group Supervisor is responsible for the management of Rapid Intervention Crew(s). The High-Rise Incident Rapid Intervention Group Supervisor's organizational responsibilities vary from the standard ICS position due to the potential for above ground operations, extended response times, and RIC(s) operating on different floors/stairwells. This position reports to the Operations Section Chief and requires close coordination with the Division/Group Supervisors and the Staging Area Manager:

a. Obtain briefing from the Operations Section Chief or Incident Commander.
b. Participate in Operations Section planning activities.
c. Determine Rapid Intervention Group needs (personnel, equipment, supplies and additional support).
d. Evaluate tactical operations in progress.
e. Evaluate floor plans, above and below emergency operations.
f. Assign and brief Rapid Intervention Crews based on number of stairwells and floors used for emergency operations.
g. Verify potential victims and hazard locations and insure that Rapid Intervention Crew(s) are prepared for possible deployment.

h. Notify Operations Section Chief or Incident Commander when Rapid Intervention Crew(s) are operational or deployed.
i. Develop Rapid Intervention Crew(s) contingency plans.
j. Secure operations and release personnel as determined by the Demobilization Plan.
k. Maintain Unit/Activity Log (ICS Form 214).

HIGH-RISE INCIDENT BASE MANAGER -The High-Rise Incident Base Manager is responsible for the management of all functions at the Base location. This position within the organization differs from the standard ICS in that a Facilities Unit is not appropriate for this type of incident and the Base Manager reports directly to the Support Branch Director (if established) or Logistics Section Chief:

a. Obtain briefing from Support Branch Director, Logistics Section Chief, or Incident Commander.
b. Participate in Support Branch/Logistics Section planning activities.
c. Determine Base needs (personnel, equipment, supplies and additional support).
d. Evaluate layout and suitability of the selected Base location:
   • Make recommendations regarding relocation, if appropriate.
e. Establish Base layout and identify functional areas to support the incident; i.e., Apparatus Parking, Crew Ready Area, Equipment Pool, Rehabilitation Area, Command Post, and Sanitation.
f. Provide for safety, security and traffic control at Base and Command Post.
g. Provide facility services at Base and Command Post; i.e., sanitation, lighting and clean up.
h. Maintain accounting of resources in Base. Periodically update Logistics Section, Planning Section or Incident Command.

i. Direct personnel and equipment to designated locations as requested.
j. Provide an auxiliary water supply to the building, if required.
k. Update Support Branch, Logistics Section or Incident Commander as directed.
l. Secure operations and release personnel as determined by the Demobilization Plan.
m. Maintain Unit/Activity Log (ICS Form 214).

HIGH-RISE INCIDENT GROUND SUPPORT UNIT LEADER

The High Rise Incident Ground Support Unit Leader is responsible for providing transportation for personnel, equipment, and supplies refilling of SCBA air cylinders; providing fueling, service and maintenance of vehicles and portable power equipment and tools; and implementing the ground level Traffic/Movement Plan at the incident including marking safe access routes and zones. The Ground Support Unit Leader reports to the Support Branch Director (if established) or the Logistics Section Chief:

a. Obtain briefing from Support Branch Director, Logistics Section Chief, or Incident Commander.
b. Participate in Support Branch/Logistics Section planning activities.
c. Identify, establish, and implement safe movement routes and exterior Safe Refuge Areas identified in the Traffic and Personnel Movement Plans.
d. Assign personnel to fueling, maintenance and support of apparatus and portable power equipment and emergency power systems as appropriate.
e. Assign personnel to SCBA air cylinder refilling, maintenance and support.
f. Maintain inventory of support and transportation vehicles, maintenance and fuel supplies.

g. Update Support Branch, Logistics Section, or Incident Commander as directed.
h. Secure operations and release personnel as determined by the Demobilization Plan.
i. Maintain Unit/Activity Log (ICS Form 214).

HIGH-RISE INCIDENT EVACUATION GROUP SUPERVISOR

The High-Rise Incident Evacuation Group Supervisor is responsible for managing the movement of building occupants through designated evacuation route(s) to a safe location. This position reports to the Operations Section Chief or Branch Director if established:

a. Obtain briefing from the Branch Director, Operations Section Chief or Incident Commander.
b. Participate in Operations Section planning activities.
c. Determine Evacuation Group requirements (personnel, equipment, supplies).
d. Ensure the evacuation in progress is to a safe location.
e. Confirm evacuation stairwell(s) with the Operations Section and Ground Support.
f. Ensure ventilation of evacuation stairwell(s) and Safe Refuge Areas.
g. Coordinate evacuation message with Systems Control Unit utilizing the building's Public Address System.
h. Assign personnel in the evacuation stairwell(s) to assist/direct building occupants to a safe location.
i. Secure operations and release personnel as determined by the Demobilization Plan.
j. Maintain Unit/Activity Log (ICS Form 214).

For more detailed information read:
High-Rise Structure Fire Operation System Description
ICS-HR-120-1

Notes

CHAPTER 20

PROTECTIVE ACTION GUIDELINES

INTRODUCTION

This section provides guidelines and procedures for protective actions when hazardous conditions develop to the degree that emergency responders must take action to protect the public at risk. Threatened or hazardous areas may be created by, but are not limited to: fires, hazardous materials, transportation accidents, floods, WMD incidents, civil disturbances, etc. Ideally, protective actions are progressive, usually initiated by alerting the public in the affected area, controlling access, sheltering in-place and finally by evacuation. However, these actions may be implemented simultaneously based on the hazard, complexity of the emergency, and the type and size of the affected area. The key to successfully conducting protective action operations is sound planning.

AUTHORITY

The decision to alert the public of a hazardous incident, restriction or closed access corridor and/or to evacuate an affected area is often made by the fire department Incident Commander. However, the authority necessary to carry out these actions usually rest with law enforcement.

ORGANIZATION

In emergency operations, there may be several lead and support agencies involved. In an incident where one agency has a preponderance of responsibility for abating the problem, a single Incident Commander from that agency shall be appointed.

In an incident where law enforcement and the fire department both have substantial responsibilities, a Unified Command organizational structure should be formed. Incident Commanders from both departments will be named. Establishing a Unified Incident Command structure better integrates incident objectives and the development of Incident Action Plans. This results in a more efficient coordination process of incident operations thereby enhancing the safety of responders and the public.

Initial Assessment and Notifications:

1. Identify hazard and risk to the public; determine the affected area and plot on a map the identified page, alphanumeric grid and quadrant of grid. Example: TB Page 689 A4 Northeast quadrant.

2. Notify jurisdictional law enforcement agency of emergency situation and recommended protective action.

3. Insure the appropriate local Emergency Services Agency is notified with regard to recommended protective actions. Ensure if evacuation is planned that evacuation centers are identified in safe areas. Note: The management of Evacuation Centers is often delegated to the local Red Cross or other non-government organizations:

a. **Evacuation warning** – The alerting of people in an affected area(s) of potential threat to life and property. An Evacuation Warning considers the probability that an area will be affected and prepares people for a potential evacuation order.

b. **Evacuation Order** – Requires the immediate movement of people out of an affected area due to an imminent threat to life (one to two hours or less).

c. **Shelter-in-place** – Advises people to stay secure at their current location. Note: Use this tactic only if the safety of citizens can be assured by remaining in place, as evacuation will cause a higher potential for loss of life.

d. **Rescue** – Emergency actions taken within the affected area to recover and remove injured or trapped citizens. Responders have specific training and personal protective equipment necessary to accomplish the mission, i.e., hazardous material spill, swift-water rescue, etc. Boundaries of the areas where rescue is planned should be identified on the incident map with notation that entry is restricted to rescue workers only. Note: Depending on the size, type and complexity of the emergency, all of the above protection actions could be employed on the same incident.

Immediate Evacuation Checklist:

a. Establish and collocate the Incident Command Post to include all cooperating agencies, i.e., law enforcement, fire, health department, local emergency management agency, etc.

b. Establish Unified Command when appropriate. Unified Incident Commanders should jointly assess and report incident potential and request adequate resources to accomplish agreed upon objectives.

c. Jointly develop the incident Evacuation Plan; ensure that the planning process is conducted under the unified command process with input from lead and support agencies as required. Keep in mind that many local jurisdictions have developed emergency evacuation plans for high hazard areas.

d. Clearly identify on a map the area(s) that are under an immediate threat and/or an evacuation order (recommend using the U.S. National Grid).

e. Clearly identify on a map the potential areas of risk based on spread of the incident. These areas may be identified as under an evacuation warning.

f. Identify evacuation routes to nearest safe location. This information will be critical for shelter locations and should be shared with the local emergency services agency, Red Cross or NGO responsible for shelter identification and management.

g. For planning purposes, approximately 2,500 autos per lane per hour can be accommodated on most roads with an average occupancy of four persons per vehicle.

h. Planning evacuations for special facility and populations will require additional time and attention to detail. These may include hospitals, elder care facilities, and the like.

i. Evacuation planning should also consider timelines, transportation needs and contacts required for large animal evacuations.

j. Identify on a map area/locations where shelter-in-place locations are designated. These areas may require verification by the Operations Section Chief and concurrence from the Incident Commander(s).

k. Determine traffic control points. Control points should be located on all sides of the incident and outside the threat area. The perimeter established for traffic control will depend on both the affected population and traffic density.

l. Traffic closure levels - Display on incident and public information maps:

    • Level 4 – closed to all traffic, potential life hazard
    • Level 3 – closed to all traffic except emergency responders
    • Level 2 – closed to all traffic except emergency responders and critical resources, i.e., public works, electrical service, animal rescue
    • Level 1 – open to above resources and residents only

m. The completed Evacuation Plan should be distributed to all command and general staff members and their subordinates. Additionally, copies should be distributed to all lead and support agencies, local elected officials and the respective county or city emergency operations centers.

Re-Entry Planning Checklist:

a. Identify re-entry date and time
b. Identify area(s) to be re-entered
c. Type of re-entry, homeowner/landowner only with identification or general public
d. Considerations:

- Is the threat mitigated?
- Are power lines secured?
- Are transportation systems hazards mitigated, i.e., roads cleared, bridges inspected, hazard trees removed, etc.?
- Incident Commanders' approval granted
- Local law enforcement agencies' approval granted
- Other local emergency service agencies' approval granted (911 service)
- Utility agency informed and supports decision
- Local EOC notified and approves

Notes

CHAPTER 21

FIREFIGHTER INCIDENT SAFETY AND
ACCOUNTABILITY GUIDELINES
ICS 910

INTRODUCTION

One of the most important issues facing the Incident Commander is personnel accountability at the scene of emergencies. These Firefighter Incident Safety and Accountability Guidelines incorporate additional firefighter safety measures and personnel accountability into the Incident Command System (ICS) to ensure compliance with NFPA standards.

The NFPA 1500 and 1561 Standards contain specific requirements regarding accountability of members that include but are not limited to the following:

Firefighter Emergencies

When Firefighters or incident personnel are faced with life threatening emergencies, they may call for help using a variety of verbiage. Incident Commanders shall acknowledge the person in trouble and use the term "EMERGENCY DOWN," "FIREFIGHTER MISSING," or "FIREFIGHTER TRAPPED," to all incident personnel.

Other guidelines for "EMERGENCY TRAFFIC" include the Dispatch Center transmitting a distinctive "EMERGENCY TRAFFIC" tone on designated channel(s) followed by clear text verbal message that identifies the type of emergency, i.e., "FIREFIGHTER DOWN," "FIREFIGHTER MISSING," or "FIREFIGHTER TRAPPED."

Rapid Intervention Crew/Company (RIC) Members

Rapid Intervention personnel have two very important duties. These are:

- Monitor designated radio channel(s) while standing by and during rescue operations.
- Initiate rescue plan assigned by the Incident Commander or Operations Section Chief.

In the initial stages of an incident where only one team is operating in the hazardous area at a working structural fire, a minimum of four individuals is required, consisting of two individuals working as a team in the hazard area and two individuals present outside this hazard area for assistance or rescue at emergency operations where entry into the danger area is required. The standby members shall be responsible for maintaining a constant awareness of the number and identity of members operating in the hazardous area, their location and function, and time of entry. The standby members shall remain in radio, visual, voice or signal-line communications with the team (NFPA 1500 6-4.4).

Members that arrive on the scene of a working structural fire prior to the assembling of four persons can initiate exterior actions in preparation for an interior attack.

Initial attack operations shall be organized to ensure that, if upon arrival at the emergency scene, initial attack personnel find an imminent life-threatening situation which immediate action could prevent the loss of life or serious injury, such action shall be permitted with less than four personnel when conducted in accordance with NFPA 1500 Section 6-2. No exception shall be permitted when there is no possibility to save lives. Any such actions taken in accordance with this section shall be thoroughly investigated by the fire department with a written report submitted to the fire chief (NFPA 1500 6-4.4.5).

In the initial stages of an incident, the IC supervises the RIC. As the incident grows in complexity, this supervision can be assigned to the Operations Section Chief or even to individual Divisions to ensure the most rapid and effective deployment on a rescue.

When sufficient personnel are on-scene, the rapid intervention capability for the incident should be raised from the two-in, two-out minimum to include an entire crew or company. In some instances, such as multiple and/or remote entrance points, multiple RIC elements should be assigned and a Rapid Intervention Group Supervisor activated to supervise positioning and deployment of these Crews/ Companies.

In high-rise fire incidents, the RIC should typically be located at Staging. This will allow for RIC's to be deployed in a timely manner. Consider multiple RIC's if multiple floors are involved with positioning based on the assigned floor.

If a RIC is deployed to provide a rescue of a firefighter, the Incident Commander shall assign an additional RIC as a backup for the RIC that was deployed. Members working in the immediate area should be notified by the Incident Commander to assist in the rescue if at all possible. The IC must remember to continue to keep sufficient forces engaged in controlling the spread of the fire if threatening the trapped, lost, or injured firefighter.

Additional Rapid Intervention Considerations

When preparing for a firefighter rescue, consider the worst-case scenario. Rapid Intervention Crew/Company (RIC) standard operating guidelines are incident driven.

Officers or members assigned the task of RIC shall not get involved in routine firefighting activities, but remain in a state of readiness keeping company members together and ready for deployment.

Operational Retreat Guidelines

In addition to radio traffic requiring evacuation, the following standardized audible signal can be used to indicate evacuation.

The **EVACUATION SIGNAL** will consist of repeated short blasts of the air horn for approximately ten seconds, followed by ten seconds of silence. This sequence of air horn blasts for ten seconds followed by a ten-second period of silence will be done three times; total air horn evacuation signal including periods of silence will last fifty seconds. This should be done in conjunction with the radio announcement of "EMERGENCY TRAFFIC," with direction for emergency scene personnel to evacuate the hazard area.

The Dispatch Center should continue to advise the Incident Commander of the elapsed time at each additional fifteen-minute interval, or until canceled by the IC, or until the incident is declared under control, i.e., knockdown.

PROCEDURES FOR THE IDENTIFICATION AND MANAGEMENT OF LIFE HAZARD ZONES

INTRODUCTION

Incident Commanders are responsible for the safety of all incident personnel and may have to take action to protect personnel from life threatening conditions that on-scene fire personnel and other responders do not have the capabilities, tools, or training to immediately mitigate. These actions may include:

- Immediate notification of personnel
- Notification for ongoing or long-term life hazards
- Methods to isolate and clearly identify the life hazard with three strands of barrier tape
- Assignment of Lookouts or Assistant Safety Officers when needed
- Identification methods for remote or large area life hazards

The clearly identifiable method to assure that fire personnel and other responders do not enter Life Hazard Zones includes the use of a minimum of three (3) horizontal strands of barrier tape that states "**Do Not Enter**" or "**Do Not Cross**," to prevent entry to the hazardous area. Three horizontal strands of any Fireline tape or flagging tape between one inch and three inches with the words "Do Not Enter" or "Do Not Cross," securely fixed to stationary supports, and in sufficient locations to isolate the hazard, will meet the requirement of identifying a Life Hazard Zone.

DEFINITIONS

Life Hazard: The existence of a process or condition that would likely cause serious injury or death to exposed persons.

Life Hazard Zones: A system of barriers surrounding designated areas at the incident scene that is intended to **STOP** fire personnel and other responders from entering a potentially Life Threatening, Hazardous Area.

Life Hazard Lookout: A qualified person in a location where they can safely observe a Life Hazard, monitor resources and personnel in the area, and communicate with resources keeping them a safe distance away. The lookout will also isolate and deny entry to any responders or resources until the life hazard is mitigated and the Incident Commander approves the release of the Life Hazard Zone.

INFORMATION AND GUIDELINES

Whenever a life hazard is present, or an immediate threat to the health and safety of incident personnel is present at an incident, any person who recognized the potential life hazard shall immediately contact the Incident Commander using **EMERGENCY TRAFFIC** to advise of the situation. Included in the Emergency Traffic notification:

- Type/Nature of the hazardous condition (i.e., downed electrical wires, imminent building collapse, etc.)
- Specific location
- Resource needs
- Any immediate exposure needs or issues

Incident Commander shall request the appropriate resource or agency to respond to the incident to evaluate and mitigate the life hazard (i.e., Utility Company, Structural Engineer, etc.) and assign a lookout or Assistant Safety Officer until Life Hazard Zone(s) is established.

The Incident Commander shall assign a life hazard lookout to prevent any incident personnel from entering the area until such time as the procedures below have been completed.

Identification of Life Hazard Zones

a. The Standard for identification of a LIFE HAZARD ZONE:

1. Deploy barrier tape in the following manner to prevent entry and identify the hazard zone. The optimal tape would be red and white striped or chevron barrier tape that states "**Life Hazard – Do Not Enter**," however, existing Fire or Police perimeter tape that includes the words "Do Not Enter" or "Do Not Cross" will meet this standard.

2. The tape shall be configured in **three horizontal strands** approximately 18 to 24 inches apart and securely fixed to stationary supports to establish the LIFE HAZARD ZONE. The LIFE HAZARD ZONE d shall be of sufficient size to provide complete isolation, distance and protection from the hazard, and supports shall be capable of supporting the barrier tape throughout the incident.

3. The use of illumination is recommended to enhance nighttime visibility to further identify the LIFE HAZARD ZONE. Examples include orange cones with a flashing strobe light on the ground, or glow sticks securely attached to the barrier tape.

b. The Established Life Hazard Zone:

1. **THE THREE HORIZONTAL STRAND CONFIGURATION OF RED AND WHITE STRIPED OR CHEVRON BARRIER TAPE SHALL <u>ONLY</u> BE USED FOR LIFE HAZARD IDENTIFICATION. WHEN INCIDENT PERSONNEL SEE THE THREE- STRAND CONFIGURATION OR BARRIER TAPE, IT SHALL BE**

RECOGNIZED AS THE STANDARD FOR ISOLATING A LIFE HAZARD, AND INCIDENT PERSONNEL <u>SHALL NOT</u> ENTER THE LIFE HAZARD ZONE.

2. Ensure the LIFE HAZARD ZONE measures provide visibility to approaching personnel to prevent entry into the area throughout the duration of the incident.

3. Maintain the LIFE HAZARD ZONE for the duration of the incident <u>or hazard</u>. Approval from the IC is required prior to the removal of the Life Hazard Zone barriers.

4. The LIFE HAZARD ZONE identification measures are intended to provide a visual cue to all incident personnel. Life Hazard Lookout(s) or Assistant Safety Officers shall be considered to ensure a physical barrier between personnel and the LIFE HAZARD ZONE through effective communications and notifications.

5. The Incident Commander shall be responsible for ensuring that all incident personnel are notified of the Life Hazard Zone. This may be accomplished through any approved method such as face-to-face, emergency traffic radio messages or the Incident Action Plan.

c. Remote Locations: In cases where the extent of the hazard zone is so large that is not practical to completely isolate the area, such as on large incidents in remote locations, the following will be the minimum standard for these situations:

1. The Incident Commander must approve the use of these minimum standards for each Life Hazard:

   The Incident Commander shall assign a life hazard lookout at appropriate access points to prevent any incident personnel from entering the area until such time as the procedures below have been completed.

   Three horizontal stripes of red and white Life Hazard tape or barrier tape (as described above) will be affixed to two vertical uprights at appropriate locations along the access route to the Life Hazard area. A description of the hazard, location of the hazard, and distance from the Life Hazard indicator tape to the hazard shall be attached at each location.

2. All personnel working in the area or Division shall be notified of the Life Hazard immediately. Incident personnel may be notified through the routine briefings, emergency traffic, radio messages, the Incident Action Plan, and the Incident Map.

3. The location(s) of the Life Hazard(s) and Placard(s) shall be marked on the Incident Map using standardized symbols. The symbol to mark the Life Hazard Zone on the incident map is a red octagon (Stop Sign) with three white horizontal lines with a description of the hazard noted underneath.

- Personnel shall not breach, alter, or remove any LIFE HAZARD ZONE identification measures until the hazard has been abated and approval granted by the Incident Commander.
- All personnel have a personal responsibility to be aware of LIFE HAZARDS and make proper notifications when they are encountered at an incident.
- Remember the slogan: **THREE STRIPES, YOU'RE OUT!**

For More Detailed Information Read:
Firefighter Incident Safety And Accountability Guidelines
ICS 910

Notes

CHAPTER 22

GLOSSARY OF TERMS

This glossary contains definitions of terms frequently used in ICS documentation that are, for the most part, not defined elsewhere in this guide.

29 CFR Part 1910.120. Item 29 of the Code of Federal Regulations, Part 1910.120 in the Hazardous Waste operations and Emergency Response reference document as required by SARA. This document covers employees involved in certain hazardous waste operations and any emergency response to incidents involving hazardous situations. Federal OSHA enforces this code.

Access Control Point. The point of entry and exit from control zones that regulate the traffic to and from the work areas and control zones.

Agency Executive or Administrator. A chief executive officer (or designee) of an agency or jurisdiction that has responsibility for the incident.

Agency Representative. An individual assigned to an incident from an assisting or cooperating agency that has been delegated authority to make decisions on matters affecting that agency's participation at the incident. Agency Representatives report to the Incident Liaison Officer.

Air Monitoring. The use of devices to detect the presence of known or unknown gases or vapors.

Air Transportable Mobile Weather Unit (ATMWU). A portable weather data collection and forecasting system used by a National Weather Service Fire Weather Forecaster.

All Risk. Any incident or event, natural or human-caused that warrants action to protect life, property, environment, public health or safety, and minimize disruption of government, social or economic activities.

ALS (Advanced Life Support). Allowable procedures and techniques utilized by EMT-P and EMT-II personnel to stabilize critically sick and injured patient(s) that exceed Basic Life Support procedures.

ALS Responder. Certified EMT-P or EMT-II.

Area Command. Area Command is an expansion of the incident command function primarily designed to manage a very large incident that has multiple incident management teams assigned. However, an Area Command can be established at any time that incidents are close enough that oversight direction is required among incident management teams to ensure conflicts do not arise.

Assigned Resources. Resources checked in and assigned work tasks on an incident.

Assistant. Title for subordinates of Command Staff positions. The title indicates a level of technical capability, qualifications, and responsibility subordinate to the primary positions. Assistants may also be used to supervise unit activities at camps.

Assisting Agency. An agency directly contributing suppression, rescue, support, or service resources to another agency.

Available Resources. Resources assigned to an incident and available for an assignment.

Base. That location where the primary logistics functions are coordinated and administered (incident name or other designator will be added to the term "Base"). The Incident Command Post may be co-located with the base. There is only one base per incident.

Basic Rope Rescue. Rescue operations of a non-complex nature employing the use of ropes and accessory equipment.

BLS (Basic Life Support). Basic non-invasive first-aid procedures and techniques utilized by EMT-P, EMT-II, EMT-I, EMT-D and First Responder personnel to stabilize sick and injured patient(s).

BLS Responder. Certified EMT-I or First Responder.

Boat drive-air. A boat with a propulsion system using an aviation propeller or a ducted fan to generate thrust from the engine having an on-plane draft of zero to twelve inches. The typical boats of this category are the "Florida Swamp" boats and surface effect boats.

Boat drive-jet. A boat with a propulsion system using a water pump to generate thrust having an on-plane draft of six to twelve inches. They can be susceptible to damage from floating debris.

Boat drive-propeller. A boat with a propulsion system using a propeller to generate thrust having an on-plane draft of eighteen to twenty-four inches.

Boat, non-powered. A non-motorized vessel capable of safely transporting rescuers or victims (e.g., raft, skiff, johnboat, etc.).

Boat, powered. A motorized vessel capable of safely transporting rescuers or victims, (e.g. IRB: "Inflatable Rescue Boat", RHIB: "Rigid Hull Inflatable Rescue Boat", Rigid Hull Boat, PWC: "Personal Water Craft," "Airboat", etc.).

Branch. That organizational level having functional, geographical, or jurisdictional responsibility for major parts of the incident operations. The Branch level is organizationally between Section and Division/Group in the Operations Section, and between Section and Units in the Logistics Section. Branches are identified by the use of Roman Numerals, by function, or jurisdictional name.

Camp. A geographical site, within the general incident area, separate from the base, equipped and staffed to provide food, water, and sanitary services to incident personnel.

Chemical Protective Clothing. Includes complete NFPA compliant ensembles (garment, gloves and boots) of individual replaceable elements (boots, gloves) designed and certified to provide protection for the wearer against the physical and chemical effects of hazardous materials.

CHEMTREC. Chemical Transportation Emergency Center operated as a public service of the Chemical Manufacturers Association.

Clear-Text. Use of plain English and common terminology understandable by all.

Command. The act of directing, ordering and/or controlling resources by virtue of explicit legal, agency, or delegated authority.

Command Staff. The Command Staff consists of the Public Information Officer, Safety Officer, and Liaison Officer who report directly to the Incident Commander.

Company Unity. A term to indicate that a fire company or unit shall remain together in a cohesive and identifiable working group, to ensure personnel accountability and the safety of all members. A company officer or unit leader shall be responsible for the adequate supervision, control, communication and safety of members of the company or unit.

Compatibility. The matching of personal protective equipment (PPE) to the hazards involved providing the best protection for the worker.

Complex. A complex is two or more individual incidents located in the same general proximity that is assigned to a single Incident Commander or Unified Command to facilitate management.

Confined Space Rescue. Rescue operations in an enclosed area, with limited access/egress, not designed for human occupancy and has the potential for physical, chemical or atmospheric injury.

Contamination Control Line (CCL). The established line that separates the Contamination Reduction Zone from the Support Zone.

Contamination Reduction Corridor (CRC). A corridor within the Contamination Reduction Zone where decontamination procedures are conducted.

Contamination Reduction Zone (CRZ). The area between the Exclusion Zone and the Support Zone that acts as a buffer to separate the contaminated area from the clean area.

Control Zones. The geographical areas within the control lines set up at a hazardous materials incident. Includes the Exclusion Zone, Contamination Reduction Zone and Support Zone.

Cooperating Agency. An agency supplying assistance other than direct suppression, rescue, support, or service functions to the incident control effort (e.g., Red Cross, telephone company, etc.).

Coordination Center. A facility that is used for the coordination of agency or jurisdictional resources in support of one or more incidents.

Cost Sharing Agreements. Agreements between agencies or jurisdictions to share designated costs related to incidents.

Decontamination (DECON). The physical and/or chemical process of removing or reducing contamination from personnel or equipment, or in some other way preventing the spread of contamination by persons and equipment.

Delayed Treatment. Second priority in patient treatment. These people require aid, but injuries are less severe.

Delegation of Authority: A statement provided to the Incident Commander by the Agency Executive delegating authority and assigning responsibility. The Delegation of Authority can include objectives, priorities, expectations, constraints, and other considerations or guidelines as needed. Many agencies require written Delegation of Authority to be given to Incident Commanders prior to their assuming command on larger incidents.

Deputy. An individual assigned to the Incident Commander, General Staff, or Branch Directors with equal qualifications and delegated authority when acting in their absence.

Division. That organization level having responsibility for operations within a defined geographic area. The Division level is organizationally between the Strike Team and the Branch (see also "Group").

Emergency Traffic. The term used to clear designated channels used at an incident to make way for important radio traffic for a firefighter emergency situation or an immediate change in tactical operations.

EMT-I (Emergency Medical Technician-I). An individual trained in Basic Life Support procedures and techniques and who has a valid EMT-I certificate.

EMT-II (Emergency Medical Technician-II). An individual with additional training in limited Advanced Life Support procedures and techniques according to prescribed standards and who has a valid EMT-II certificate.

EMT-D (Emergency Medical Technician-Defibrillator). An Emergency Medical Technician I with training and certification in automatic and semi-automatic external defibrillation.

EMT-P (Emergency Medical Technician-Paramedic). An EMT-I or EMT-II who has received additional training in Advanced Life Support procedures and techniques and who has a valid EMT-P certificate or license.

Environmental. Atmospheric, Hydrologic and Geologic media (air, water and soil).

Evacuation: Organized, phased, and supervised withdrawal, dispersal, or removal of civilians from dangerous or potentially dangerous areas, and their reception and care in safe areas.

Exclusion Zone (EZ). The innermost area immediately surrounding a hazardous materials incident that corresponds with the highest degree of known or potential hazard, and where entry may require special protection.

Expanded Medical Emergency. Any medical emergency that exceeds normal first response capabilities.

Field Testing. The identification of chemical substances using a variety of sources and testing kits that assist in identifying associated chemical and physical properties of those tested chemicals.

Fireline Emergency Medical Technician (FEMT). The FEMT provides emergency medical care to personnel operating on the fireline.

Flood Evacuation Boat (FEB). Resource with personnel trained to operate in floodwaters with the specific task of evacuating persons or small domestic animals from isolated areas.

General Staff. The group of incident management personnel comprised of the Operations Section Chief, Planning Section Chief, Logistics Section Chief, and Finance/Administration Section Chief.

Group. Groups are established to divide the incident into functional areas of operation. Groups are located between Branches (when activated) and Resources in the Operations Section. (See Division).

Hazardous Material. Any solid, liquid, gas, or mixture thereof that can potentially cause harm to the human body through respiration, ingestion, skin absorption or contact and may pose a substantial threat to life, the environment, or to property.

Hazardous Materials Categorization. A process to determine hazardous materials classification, and chemical and physical properties of unknown substances.

Hazardous Materials Categorization Test (HAZ CAT). A field analysis to determine the hazardous characteristics of an unknown material.

Hazardous Materials Company. Any piece(s) of equipment having the capabilities, PPE, equipment, and complement of personnel as specified in the Hazardous Materials Company Types and Minimum Standards found in the Field Operations Guide (ICS 420-1).

Hazardous Materials Incident. The uncontrolled release or threat of release of a hazardous material that may impact life, the environment, or property.

Heavy Floor Construction. Structures of this type are built utilizing cast-in-place concrete construction consisting of flat slab panel, waffle or two-way concrete slab assemblies. Pre-tensioned or post-tensioned reinforcing steel rebar or cable systems are common components for structural integrity. The vertical structural supports include integrated concrete columns, concrete enclosed or steel frame, that carry the load of all floor and roof assemblies. This type includes heavy timber construction that may use steel rods for reinforcing. Examples of this type of construction include offices, schools, apartments, hospitals, parking structures and multi-purpose facilities. Common heights vary from single-story to high-rise structures.

Heavy Wall Construction. Materials used for construction are generally heavy and utilize an interdependent structural or monolithic system. These types of materials and their assemblies tend to make the structural system inherently rigid. This construction type is usually built without a skeletal structural frame. It utilizes a heavy wall support and assembly system to provide support for the floors and roof assemblies. Occupancies utilizing tilt-up concrete construction are typically one to three stories in height and consist of multiple monolithic concrete wall panel assemblies. They also use an interdependent girder, column and beam system for providing lateral wall support of floor and roof assemblies. Occupancies typically include commercial, mercantile and industrial. Other examples of this type of construction include reinforced and un-reinforced masonry (URM) buildings typically of low-rise construction, one to six stories in height, and of any type of occupancy.

Helibase. A location within the general incident area for parking, fueling, maintenance, and loading of helicopters.

Helicopter Rescue Operational. Personnel trained and equipped to work with helicopters and crew, for hoist, short haul-line victim extraction, rappel, or low-level insertions.

Helispot. A location where a helicopter can take off and land.

Helitanker. A helicopter equipped with a fixed tank, Air Tanker Board Certified, capable of delivering a minimum of 1,000 gallons of water, retardant, or foam.

Hospital Alert System. A communications system between medical facilities and on-incident medical personnel that provides available hospital patient receiving capability and/or medical control.

Immediate Treatment. A patient who requires rapid assessment and medical intervention for survival.

Incident Action Plan (IAP). A plan that contains SMART objectives that reflects the incident strategy and specific control actions for the current or next operational period.

Incident Command Post (ICP). That location at which the primary command functions are executed and usually collocated with the incident base.

Incident Command System (ICS). The combination of facilities, equipment, personnel, procedures, and communications operating within a common organizational structure with responsibility for the management of assigned resources to effectively accomplish stated objectives pertaining to an incident.

Incident Objectives. Statements of guidance and direction that are specific, measurable, attainable, results-oriented, timely, and necessary for the selection of appropriate strategy(ies), and the tactical direction of resources.

Infrared (IR). A heat detection system used for fire detection, mapping and hot spot identification.

Infrared (IR) Groundlink. A capability through the use of a special mobile ground station to receive air-to-ground infrared imagery for interpretation.

Initial Response. Resources initially committed to an incident.

IRB. Inflatable rescue boat.

Joint Information System (JIS): Integrates incident information and public affairs into a cohesive organization designed to provide consistent, coordinated, timely information during crisis or incident operations. The mission of the JIS is to provide a structure and system for developing and delivering coordinated inter-agency messages; developing, recommending, and executing public information plans and strategies on behalf of the IC; advising the IC concerning public affairs issues that could affect a response effort; and controlling rumors and inaccurate information that could undermine public confidence in the emergency response effort.

Jurisdictional Agency. The agency having responsibility for a specific geographical area or function as designated by statute or contract.

Light Frame Construction. Materials used for construction are generally lightweight and provide a high degree of structural flexibility to applied forces, such as earthquakes, hurricanes, tornadoes, etc. These structures are typically constructed with a skeletal structural frame system of wood or light gauge steel components, which provide support to the floor or roof assemblies. Examples of this construction type are wood frame structures used for residential, multiple low-rise occupancies and light commercial occupancies up to four stories in height. Light gauge steel frame buildings include commercial business and light manufacturing occupancies and facilities.

Medical Supply Cache. A cache consists of standardized medical supplies and equipment stored in a predetermined location for dispatch to incidents.

Message Center. The Message Center receives, records, and routes information about resources reporting to the incident, resource status, and administration and tactical traffic.

MICU (Mobile Intensive Care Unit). Refers to a vehicle equipped to support paramedic functions. It would include drugs, medications, cardiac monitors and telemetry, and other specialized emergency medical equipment.

Minor Treatment. These patients' injuries require simple rudimentary first-aid.

Mobilization Center. An off-incident location at which emergency service personnel and equipment are temporarily located pending assignment, release, or reassignment.

Morgue (Temporary On-Incident). Area designated for temporary placement of the dead.

Multi-Agency Coordination (MAC). The coordination of assisting agency resources and support to emergency operations.

Multi-Agency Coordination System (MACS). The combination of facilities, equipment, personnel, procedures, and communications integrated into a common system with responsibility for coordination of assisting agency resources and support to agency emergency operations.

Multi-Casualty. The combination of numbers of injured personnel and type of injuries that exceed the capability of an agency's normal first response.

Operational Period. The period of time scheduled for execution of a given set of tactical actions as specified in the Incident Action Plan.

Operations Coordination Center (OCC). The primary facility of the Multi-Agency Coordination System. It houses the staff and equipment necessary to perform the MACS functions.

Orthophoto Maps. Aerial photographs corrected to scale so that geographic measurements may be taken directly from the prints.

Out-of-Service Resources. Resources assigned to an incident but unable to respond for mechanical, rest, or personnel reasons.

Patient Transportation Recorder. Responsible for recording pertinent information regarding off-incident transportation of patients.

Personal Protective Equipment (PPE). That equipment and clothing required to shield and/or isolate personnel from thermal, chemical, radiological, physical, or biological hazards.

Personnel Accountability. The ability to account for the location and status of personnel.

Personnel Accountability Reports (PAR). Periodic reports verifying the status of responders assigned to an incident.

PFD. Personal flotation device with a minimum U.S. Coast Guard rating of Type III or V.

Planning Meeting. A meeting, held as needed throughout the duration of an incident, to select specific strategies and tactics for incident control operations and for service and support planning.

Pre-Cast Construction. Structures of this type are built utilizing modular pre-cast concrete components that include floors, walls, columns and other sub-components that are field connected upon placement on site. Individual concrete components utilize imbedded steel reinforcing rods and welded wire mesh for structural integrity and may have either steel beam, column, or concrete framing systems utilized for the overall structural assembly and building enclosure. These structures rely on single or multi-point connections for floor and wall enclosure assembly and are a safety and operational concern during collapse operations. Examples of this type of construction include commercial, mercantile, office and multi-use or multi-function structures including parking structures and large occupancy facilities.

Protective Actions. The actions taken to preserve the health and safety of emergency responders and the public during an incident involving releases of hazardous materials. Examples would include evacuations or in-place protection techniques.

PWC. Personal watercraft (water bike, jet ski).

Qualified. A person meeting a recognized level of training, experience and certification for the assigned position.

Radiation Monitoring and Detection. The use of specialized devices to determine the presence, type and intensity of ionizing radiation, and to determine dosage over time.

Radio Cache. A cache may consist of a number of portable radios, a base station and, in some cases, a repeater stored in a predetermined location for dispatch to incidents.

Rapid Intervention Crew/Company (RIC). A crew or company designated to standby in a state of readiness to rescue emergency personnel.

Refuge Area. An area identified within the incident for the assembly of individuals in order to reduce the risk of further contamination or injury.

Reinforced Response. Those resources requested in addition to the initial response.

Reporting Locations. Any one of six facilities/locations where incident assigned resources may check in.

Resources. All personnel and major items of equipment available, or potentially available, for assignment to incident tasks on which status is maintained.

Respiratory Protection. The provision of a NIOSH approved breathing system to protect the respiratory system of the wearer from hazardous atmospheres.

Responder Rehabilitation. The rest and treatment of incident personnel who are suffering from the effects of strenuous work and/or extreme conditions.

RHIB. Rigid hull inflatable boat.

Rigid Hull. A boat constructed of wood, fiberglass, or aluminum with no inflated components.

Safe Refuge Area (SRA). A safe area within the Contamination Reduction Zone (CRZ) for the assembly of individuals who were on site at the time of the spill. Separation of any potentially contaminated or exposed persons from non-exposed persons should be accomplished in the SRA.

Search Marking System. A standardized marking system employed during and after the search of a structure for potential victims.

SEAT. Single Engine Airtanker.

Section. The organization level having functional responsibility for primary segments of incident management (Operations, Planning, Logistics, Finance/Administration). The Section level is organizationally between Branch and Incident Commander.

Single Resource. An individual piece of equipment and its personnel complement, or an established crew or team of individuals with an identified work supervisor that can be used on an incident.

Site. That area within the Contamination Reduction Control Line at a hazardous materials incident.

Site Safety and Control Plan (ICS 208). An emergency response plan describing the general safety procedures to be followed at an incident involving hazardous materials, and prepared in accordance with CCR Title 8, Section 5192, and 29 CFR 1910.120.

SMART - S.M.A.R.T. Acronym for Specific, Measurable, Attainable, Results oriented, and Timely.

Staging Area. That location where incident personnel and equipment are assigned on a three-minute available status.

Standby Members (2-in, 2-out). Two personnel who remain outside the hazard area during the initial stages of an incident to rescue responders and who are responsible for maintaining a constant awareness of the number and identity of members operating in the hazardous area, their location and function, and time of entry.

START - S.T.A.R.T. Acronym for Simple Triage And Rapid Treatment.

Strategy. The general plan or direction selected to accomplish incident objectives.

Strike Team. Specified combinations of the same kind and type of resources, with common communications and a leader.

Structure/Hazards Marking System. A standardized marking system to identify structures in a specific area and any hazards found within or near the structure.

Support Zone. The area outside of the Contamination Control Line where equipment and personnel are assembled in support of incident operations, wherein such personnel and equipment are not expected to become contaminated.

Swiftwater. Water that is moving fast enough to produce sufficient force to present a significant life and safety hazard to a person entering the water.

Training Levels:

Awareness: Knowledge based course of instruction, emphasizing hazards and personnel safety. Generally lecture only.

Operational: Participation based course of instruction; emphasizing personal safety, team safety and limited low risk victim rescue. The course generally includes objective evaluation and testing.

Technician: Performance based course of instruction emphasizing personnel safety, team safety, and mid to high-risk victim rescue. The course generally includes objective evaluation and testing.

Tactics. Deploying and directing resources on an incident to accomplish the objectives designated by current incident strategy.

Task Force. A group of resources with common communications and a leader that may be pre-established and sent to an incident, or formed at an incident.

Technical Reference. Access to, use of, and interpretation of various technical databases, chemical substance data depositories, response guidelines, regulatory documents, and other sources both in print and electronic format.

Technical Specialists. Personnel with special skills who are activated only when needed.

Triage. Screening and classification to determine priority needs in order to ensure the efficient use of personnel, equipment and facilities.

Triage Tag (medical). A tag used by triage personnel to identify and document the patient's medical condition.

Unified Command. Unified Command is a team effort that allows all agencies with jurisdictional responsibility for the incident, either geographical or functional, to manage an incident by establishing a common set of incident objectives and strategies. This is accomplished without losing or abdicating agency authority, responsibility or accountability.

Unit. An organizational element having responsibility for a specific function within the Operations, Plans, Logistics, or Finance Sections.

Urban Search and Rescue (US&R) Company. Any ground vehicle(s) providing a specified level of US&R operational capability, rescue equipment, and personnel.

Urban Search and Rescue (US&R) Crew. A pre-determined number of individuals who are supervised, organized and trained principally for a specified level of US&R operational capability. They respond without equipment and are used to relieve or increase the number of US&R personnel at the incident.

Watershed Rehabilitation. Restoration of watershed to, as near as possible, its pre-incident condition, or to a condition where it can recover on its own. Also known as "rehab".

Weapons of Mass Destruction (WMD). Reference to those substances that can be weaponized and are developed for the purpose of creating widespread injury, illness and death. Agents are produced in quantity and/or filled into munitions in a specialized formulation with enhanced shelf life or dissemination properties.

Notes

WATCH OUT SITUATIONS

1. Fire not scouted and sized up.

2. In country not seen in daylight.

3. Safety zones and escape routes not identified.

4. Unfamiliar with weather and local factors influencing fire behavior.

5. Uninformed on strategy, tactics, and hazards.

6. Instructions and assignments not clear.

7. No communication link with crew members or supervisor.

8. Constructing line without safe anchor point.

9. Building fire line downhill with fire below.

10. Attempting frontal assault on fire.

11. Unburned fuel between you and fire.

12. Cannot see main fire, not in contact with someone who can.

13. On a hillside where rolling material can ignite fuel below.

14. Weather becoming hotter and drier.

15. Wind increases and/or changes direction.

16. Getting frequent spot fires across line.

17. Terrain and fuels make escape to safety zones difficult.

18. Taking nap near fireline.

www.ingramcontent.com/pod-product-compliance
Lightning Source LLC
Chambersburg PA
CBHW051441170526
45166CB00001B/69